CHANGES IN THE WIND

by Margery and Howard Facklam

FROM CELL TO CLONE: The Story of Genetic Engineering
THE BRAIN: Magnificent Mind Machine

by Margery Facklam

FROZEN SNAKES AND DINOSAUR BONES:
Exploring a Natural History Museum
WILD ANIMALS, GENTLE WOMEN

Changes

EARTH'S SHIFTING CLIMATE

BY MARGERY AND HOWARD FACKLAM

ILLUSTRATED WITH PHOTOGRAPHS
AND WITH DIAGRAMS BY PAUL FACKLAM

HARCOURT BRACE JOVANOVICH, PUBLISHERS
SAN DIEGO NEW YORK LONDON

The authors and the publisher wish to thank *The New York Times* for permission to quote on page 112–113 material from Lewis Thomas's article "Man's Role on Earth" that appeared on April 1, 1984. Copyright © 1984 by The New York Times Company. Reprinted by permission; and W. H. Freeman and Company for permission to quote on page 108 a paragraph from "The Climatic Effects of Nuclear War" by Turco, Toon, Ackerman, Pollack, and Sagan, *Scientific American*, August 1984.

Design by Barbara DuPree Knowles
Printed in the United States of America

Library of Congress Cataloging in Publication Data
Facklam, Margery.
Changes in the wind.
Bibliography: p.
Includes index.
Summary: Examines the factors causing changes in the earth's climate, including ocean currents, the destruction of the rain forests, and the greenhouse effect, and discusses predictions for the future.
1. Climatic changes—Juvenile literature.
[1. Climatic changes. 2. Climatology] I. Facklam, Howard. II. Title.
QC981.8.C5F33 1985 551.6 85-5475
ISBN 0-15-216115-5

A B C D E
First edition

For Christopher Matthew Facklam

CONTENTS

PREFACE

Thomas Jefferson kept a journal of weather observations, and in 1816 he wrote, ". . . the climates of the several states of our union have undergone a sensible change since the dates of their first settlements" and that since the time of Augustus Caesar the climate of Italy had "changed regularly at the rate of 1° of Fahrenheit's thermometer for every century."

"May we not hope," he wrote, "that the methods invented in the latter times for measuring with accuracy the degrees of heat and cold, and the observations which have been and will be made and preserved, will at length ascertain this curious fact in physical history?"

When we began to research the story of climate change, we found that not only had Jefferson's hopes been realized, but also that in the search for the reasons for changes in earth's climate, there is an unprecedented crossover and cooperation in the sciences.

Twenty-five years ago the first weather satellite was launched by the United States, and since then many other kinds of extremely accurate equipment and techniques have been designed to collect and measure weather changes. The data they collect are used by climatologists and meteorologists, who work with biologists, geologists, physicists, astronomers, agricultural engineers, and oceanographers.

Everyone's looking for the same things . . . a way to understand all the interlocking systems on this planet and ways for mankind to occupy a comfortable, cooperative niche as these systems change.

Margery and Howard Facklam

FEBRUARY 1985

ACKNOWLEDGMENTS

As in any book in which the authors are only the "explainers," we turned to the experts for advice and facts. Our appreciation is endless for the people who were so generous with their valuable time and expertise.

First we wish to thank our editor, Anna Bier, who suggested this book to us and then edited it with such care.

We extend special gratitude to Dr. Richard T. Barber, biologist at Duke University's Marine Lab, and his assistant and Senior Scientist aboard the research ship *Wecoma*, Jane Kogelschatz. They made it possible for us to see El Niño's effects firsthand. Other scientists who helped us are: Dr. Douglas Paine and Dr. William Pardee, Cornell University; Dr. Merle H. Jensen, University of Arizona; Astronomer and Director of the Buffalo Museum of Science, Ernest Both; State University of New York at Buffalo professors Dr. Chester Langway, Dr. Peter Caulkin, Dr. Paul Reitan, and especially Dr. Dennis S. Hodge who reviewed the manuscript.

CHANGES IN THE WIND

1

THE CHICKEN LITTLE PREDICTION

Looking like a giant glass bead, the one-man scouting submarine darted around the coral reef, brushing over the tangles of orange anemone and sweeping through beds of purple fans. A mass of silvery fish, as close as scales on a snake, swept past, responding as one to the movement of the school's leader.

"Alpha sub reporting." The pilot touched the control panel, and the small craft plunged toward the ocean floor. "On my way to desalination station one," he said. "Passing eastside kelp bed. Eight robots functioning at full power." He let the sub hover for a moment while he watched the harvesters cut great swaths through the kelp beds with their lasers. Other teams of robots operated the cranes that hauled the nets to the surface. "Looks like a rich crop," the pilot reported to the fleet captain. "On my way to Trade Towers station, passing over Long Island reef now."

The bubble-sub glided through the shallow waters covering the reef that formed a natural barrier around the specialized fish farms. He maneuvered his ship down straight lanes that separated the coral-encrusted towers. They reminded him of grotesque skeletons of buildings in a surrealistic city. He had to remind himself that it had been a city of the twentieth century. He shivered as though the ghosts of people from that ancient era still haunted the ruins of this once dynamic city called New York.

Quickly he checked the lines for the newly seeded strings of oysters that hung from the twisted steel beams. His inspection completed, the pilot threaded the sub through the remains of streets and buildings, back to the mother ship that patrolled the national fish farms. "Amazing," he muttered to himself. "Always amazing. Wonder what happened to the people? Did they move inland? Did they know the earth was warming and icecaps were melting? I wonder?"

Will it happen? Is this science-fiction scenario farfetched, or is it possible? In a few hundred years will the Empire State Building and the World Trade Center towers be barely visible above the water like great, coral-crusted piers with only marine life as tenants? Will the "greenhouse effect" warm the planet so much that polar icecaps will melt, allowing the seas to creep over coastal lands?

Is it likely that the Tower of London will disappear as the sea rushes up the Thames, or that Key West will become merely another ocean reef? Are we in danger of losing the beaches of Hawaii, Florida, and California? If the atmosphere is warming, is there anything we can do about it?

Or is the earth cooling? Are we about to move into another ice age as some scientists predict? Is it possible that the greenhouse effect will cancel out an ice age?

Are scientists and environmentalists reacting like the story book character Chicken Little who raced around the barnyard crying, "The sky is falling, the sky is falling!"?

Strange things are happening to our weather. We've had horrible heat waves, devastating droughts, and unprecedented blizzards. The normally uneventful warm ocean current called El Niño seemed to go crazy in 1982, changing weather patterns around the world. Volcanoes erupted and in a day or two left the air thicker with ash and pollutants than any number of factories spew out in a year. Acid rain continues to etch its way into land and water, and aerosols threaten to change the blanket of atmosphere.

We tend to think that climate is forever, as unchanging as the rigid diagrams in schoolbooks outlining the tropics, the temperate zones, and Arctic regions. These remembered images tell us that

climate is constant, that it is little more than an average of temperature, sunshine, and rainfall.

We've relied on that constancy. Construction engineers are supplied with statistics for the strongest gusts of wind, the extremes of temperature, the deepest frost, or the greatest flood that might be expected perhaps once in fifty or a hundred years. Farmers plant crops or raise cattle based on twenty or thirty years of meteorological observation, and nations thrive or starve depending on the climate for agriculture.

Weather is the result of a world-wide swirl of energy. It is a great web of interaction; therefore, what happens in India one week affects the price of corn on the cob a few weeks later in Indiana. If the monsoons are late in the Far East, a cold snap may freeze out the beans in Ohio. If the ocean currents average two degees warmer one winter, we can't afford anchovies on our pizza the following summer. If a volcano erupts in Mexico, the wheat crop may fail in Europe.

If weather and climate changes were merely a matter of buying a warmer parka to wear in Ohio in winter, or snow tires to put on the cars of Georgia, or umbrellas to use in the desert, the problem would be only an inconvenience. It's more than that. It's more than getting used to digging out of snowdrifts on Easter Sunday in Wisconsin or investing in irrigation systems for normally rain-drenched farms.

What's at stake is life itself. Some experts predict world-wide famine, and while others are not as pessimistic, most agree that changes in our climate will cut across political and economic boundaries. When the Environmental Protection Agency published its report in 1983 about the warming of the planet, the reactions ranged from scare headlines like "Oceans Rising As Icecaps Melt" to total disbelief.

Mankind is altering the climate in a variety of ways, but the earth itself is constantly changing as well. Just as weather is a world-wide swirl of energy, climate is also a world-wide system of interactions. And just as man's pollutants affect the atmosphere, so do a combination of natural phenomenon . . . the oceans, which cover two-thirds of the planet's surface; the great stretches of forests; the expanses of white ice and snow; volcanoes; and the engine that drives it all, the sun. Even the movement of the continents has long-range effects on climate. Measurements from two radio telescopes aloft in NASA satellites show that the North American continent is drifting toward Europe at the rate of 0.6 inches each year, and Australia is racing toward Hawaii at 2.7 inches annually.

Chicken Little's cry does not apply. The sky is not falling, but it is forever shifting. For the most part, there is little we can do to alter the course of the natural forces, but it's important to understand what is happening. Food and a place to live are still our basic needs. No longer can we fold our tents like nomads and move on to look for fresh water, a reliable food supply, and a better place for shelter. There is political and economic fallout to changes in growing seasons and to changes in fuel and water supplies caused by changes in climate.

If the storybook hen could be heard now, maybe Chicken Little would be crying, "Plan ahead, plan ahead."

2

THE WEATHER MACHINE

One hundred and eleven B-29 bombers flew in formation toward Japan at dawn. It was November 24, 1944, and their mission, with the code name San Antonio, was to bomb industrial sites near Tokyo. From a base in the South Seas, the planes droned through the sky at altitudes between 27,000 and 30,000 feet, intending to drop the bombs on a path from west to east. As they made a turn toward the east to approach Tokyo, the heavy aircraft were suddenly pushed ahead in 150-mile-an-hour winds. Neither the aircraft nor the released bombs could adjust to the wind drift. Most of the bombs missed their targets.

Other Air Force crews also had problems at high altitudes. Some reported flying west into such strong head winds that the planes seemed to stand still. With engines on full throttle, it was as though they were going nowhere. As soon as the aircraft dropped a few miles in altitude, they lost the winds. It wasn't until after the war that meteorologists found out what was happening. The Japanese already knew. They had discovered the jet stream and made good use of it.

About the same time American bombers were reporting the strong high-altitude winds, some loggers in Montana found a huge balloon stuck in a tall pine tree. It was marked with the Japanese emblem of the rising sun. Bits and pieces of similar balloons were found in the western states, and speculation ran high that the balloons carried spies or deadly germs. But the biggest worry centered around where the balloons had come from. With what we then knew of winds, it seemed impossible that the balloons could have floated all the way from Japan. They must have been launched from off-shore submarines.

It turned out, of course, that the balloons had ridden the jet stream 5,000 miles to America. Designed to drop incendiary bombs and then self-destruct to destroy the instruments, the balloons failed only because the equipment froze in the jet stream's subzero temperatures.

Like a tunnel of wind, the jet stream flows in an undulating series of loops at the outer edge of the troposphere, where it blends into the stratosphere. Wherever there are extreme temperature differences, the jet stream is fastest. When cold polar air penetrates deep into the warmer air to the south, the wintertime jet stream at the leading edge of such a cold air mass forms a ribbon 4 miles deep and 300 miles wide, traveling more than 300 miles an hour. In summer, when the temperature differences between air masses are less, the jet stream slows.

Today we make good use of this high-air current in plotting air traffic and in predicting weather. Pushed by the jet stream, we can fly faster from California to New York than the other way. On every television weathercast, we watch the jet stream's snakelike path loop southward in what the weatherman calls a trough, and we know it brings with it cool air from the north. When it heads north in a ridge, it leaves us with warmer weather.

The number of loops or waves in a jet stream varies. Meteorologists think that during the warm medieval years, it streamed across the Northern Hemisphere in four big loops, but during the Little Ice Age, the jet pattern moved south in five loops. Most of the time these loops move quickly in short periods of changing weather, but now and then they stall. Then we're in for long periods of the same kind of weather. That's what happened during the winter of 1976 and 1977, when the Northeast was held in the cold grip of blizzards and the West Coast had a long hot spell.

There are seven jet streams undulating around the earth at altitudes ranging from six to thirty miles, but they don't all occur at the same time. The polar-front jet is the one that interfered with the Air Force bombers and the one that changes weather across North America. Other jet streams in both the Southern and Northern hemispheres shift with the temperatures and the seasons.

The secrets of other wind patterns were discovered by voyagers in other ages. Columbus was looking for a route to the Indies, but the steady trade winds blew him to America. Ships carrying cargoes of horses to the New World hit low-pressure belts that becalmed them for weeks. Food and water were conserved for the crew. When the horses died of thirst and starvation, their carcasses were thrown overboard, and this region 25 to 35 degrees from the equator was tagged forever as the horse latitudes.

The forces that create these winds are part of the entire weather machine, and that machine has four basic parts:

1. The sun, the source of all the energy
2. The earth and its motion
3. The blanket of atmosphere, which is 5,600 trillion tons of gases and moisture
4. The landforms: mountains, valleys, oceans, icecaps, deserts, forests, and volcanoes.

Only a billionth of the sun's energy falls on the earth, but it is enough to keep us warm and supply the energy to sustain life. Thirty miles from the earth's surface, some of the solar energy bounces off water droplets, microscopic dust particles, and molecules of air. Almost half the incoming radiation is reflected back. Another 15 percent of the energy is absorbed by microscopic dust, water vapor, carbon dioxide, and ozone in the upper layer of atmosphere called the stratosphere. The ozone layer prevents the full force of the wavelength of light called ultraviolet from reaching us directly.

The next layer of air, called the troposphere, is about ten miles deep at the equator, tapering down to a depth of five miles at the poles. This is the layer of weather. Here, some of the sun's radiation is scattered by dust particles and moisture, and it is this scattering of light that we see as colors. Because more blue light is reflected than any other, we see the sky as blue. When the atmosphere is filled with dust or volcanic debris, different light waves are reflected, and we see brilliant red sunsets.

Because the earth is tilted, the equator gets two and a half times more sunlight per square mile than the polar regions. This unequal heating is what makes the air move. Hot air along the equator rises. Some flows north; some flows south. If the earth were smooth and unmoving, the hot air would get to the poles, cool off, descend, and flow back toward the equator to be heated again. It would be a simple, unchanging weather pattern.

Winds are the atmosphere in motion. Warm air expands and rises. Cooled air sinks. Like a large conveyor belt, air moves up from the equator, loops up toward the North Pole and down toward the South Pole, curving back again toward the equator in a series of rotating cells or belts.

The spinning of the earth deflects this continuous stream of air, so instead of blowing straight from the equator to the poles and back again, it veers off at an angle.

In the 1830s, a French physicist, Gaspard de Coriolis, was studying motion on spinning surfaces. He saw that an object moving across a turning surface spins off to the right or left, depending on the direction of the rotation. The earth spins toward the east; as a result, air and water in the Northern Hemisphere veer to the right, and below the equator they move to the left. This Coriolis force causes winds blowing north and south from the equator to become the prevailing easterlies or westerlies. In the jargon of meteorologists, easterlies mean the winds flow from the east while westerlies flow from the west.

As the sun is the furnace for the weather machine, winds are the fans that do the work. They can carry 500 million tons of topsoil in a single storm from Nebraska to South Dakota, or lift five-and-a-half-billion gallons of water per hour from the Gulf of Mexico and carry it northeast to drop as millions of tons of rain on New York. Although the working of winds is extremely complex, it is orderly. Whether it's a calm breeze, a squall over a lake, or a raging

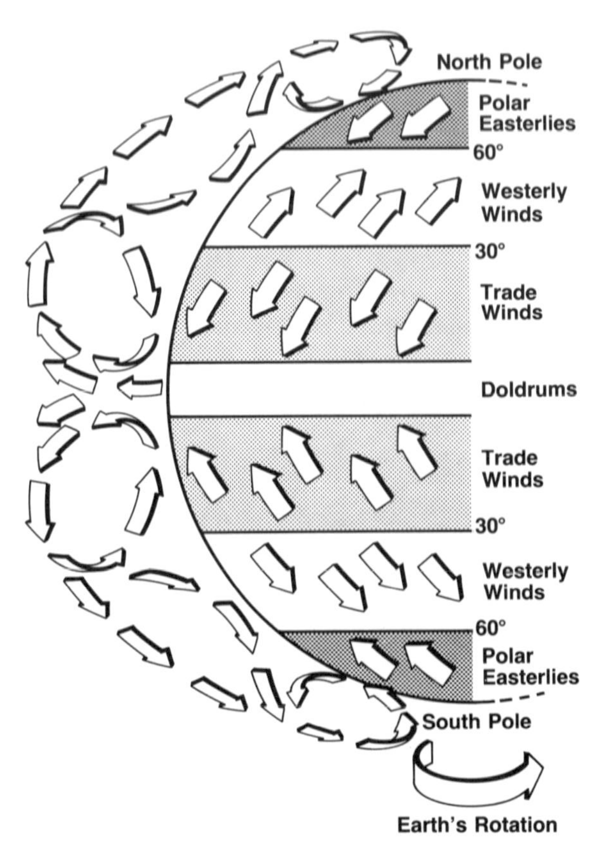

Air is set in motion in major wind belts by heat from the sun and the rotation of the earth.

hurricane, the rivers of wind are patterned in an endless interconnected system.

The earth is not smooth. Mountains and valleys interrupt and change the direction of winds and rain; volcanoes erupt and fill the stratosphere with dust that blocks the sun and cools the earth; deserts and ice fields add to the albedo effect. Albedo, from the Latin word *albus*, means white. The sun reflects off these white surfaces, changing the temperature and moisture in the air. Forests add moisture, creating clouds and rain; and the great expanses of ocean cause the biggest weather changes of all as the water absorbs the heat and stores it.

Our perception of the earth has changed since we've seen our planet from space. On April 1, 1960, the United States sent aloft its first weather satellite, TIROS-1, with two television cameras to transmit pictures of the earth's cloud cover. Since the meteorological satellite ESSA-1 was launched in February of 1966, the earth has been under constant surveillance, and more than forty environmental satellites have been put into orbit since then.

This diagram shows some of the observation satellites now circling the earth collecting environmental information from the atmosphere, the oceans, and the land. (NOAA)

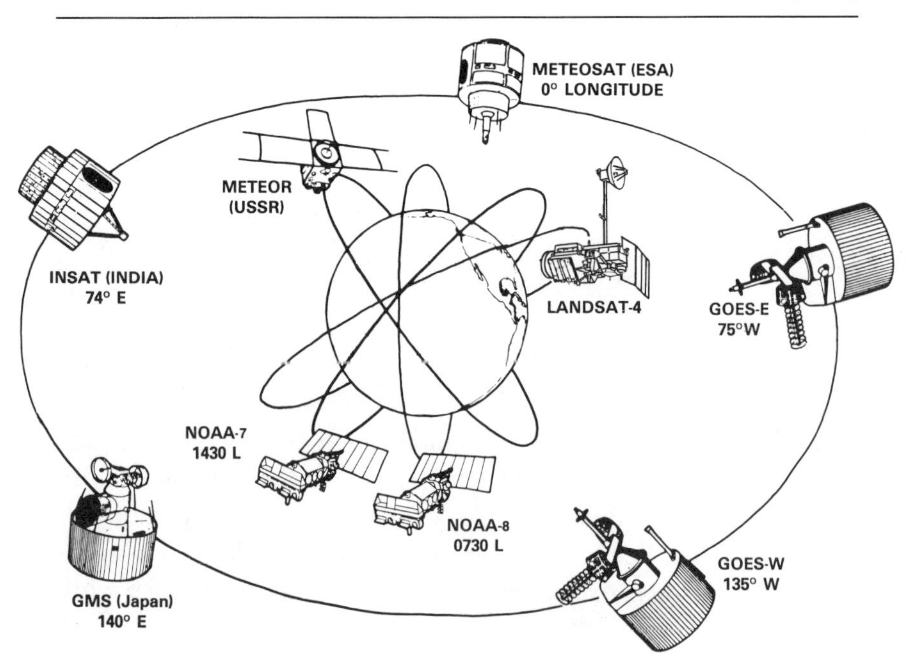

Weather information is collected by the GOES satellite system from instruments at automatic weather stations, bouys, ships, and other equipment around the world. (NOAA)

Those early satellites were crude compared to those used now. The newest satellites are equipped with sensors that can detect subtle differences in the temperature of surface water or take infrared photos that show the difference between a crop of beets or a fallow field.

The National Oceanic and Atmospheric Administration (NOAA) gathers world-wide environmental data about space, the sun, oceans, land masses, and air with satellites such as Nimbus, Landsat, Seasat, and Geosat, and they share that information with other spacefaring nations.

The view from space: Africa and southern Europe under a swirl of cloud cover were photographed by satellite. (NOAA)

◀ *The satellite GOES East sends back information on cloud cover. This photograph was relayed from a mile above the earth's surface and superimposed on an outline of the United States.* (NOAA)

(Below and on opposite page) *Engineers tested the most advanced of a series of TIROS weather satellites before packing it for shipping to Vandenberg Air Force Base in California, where it was launched in November 1984. In addition to weather instruments, the TIROS-N is equipped with search and rescue instruments that enable it to locate downed aircraft and ships in distress.* (NOAA)

On November 8, 1984, the weather satellite TIROS-N was put into orbit as part of the global weather-watching system that will also help in international search and rescue missions. TIROS-N picks up information from several hundred collection points on land, in the air, and from bouys and ships at sea. It measures temperature and moisture and keeps track of energy particles from the sun. The information gathered is used to forecast weather, to track hurricanes, and assess data for agriculture, commercial fishing, forestry, and other industries.

Why is it then that with all this sophisticated equipment, the weatherman is so often wrong?

The National Weather Service has 243 offices throughout the country with 52 national forecast stations. Thousands of dedicated meteorologists, like those who live on top of Mount Washington

◀ *The observatory and weather station on Mount Washington in Gorham, New Hampshire, is built at the highest point in the northeastern United States, 6,288 feet. It is a unique site because it is subject to extremes of climatic, meteorological, biological, and ecological stress not found elsewhere outside deep polar regions. The highest wind velocity ever recorded anywhere in the world was measured there at 231 miles per hour on April 12, 1934.*

(MOUNT WASHINGTON OBSERVATORY)

The observatory and weather station covered in ice in May 1984

(MOUNT WASHINGTON OBSERVATORY)

in New Hampshire, keep meticulous records of daily and even hourly observations. At the Weather Service, the staff studies the incoming charts and statistics, but they still can't pinpoint storms precisely because the forecasts are made for a ten-thousand-square-mile area. Because winds change rapidly in such a wide area, a forecaster can predict only within a general range. If the prediction is for a 10 percent chance of showers, there's a 90 percent chance you won't need your umbrella.

With the new unit called PROFS (Program for Regional Observing and Forecasting Services), however, it will be possible to locate severe storms six hours in advance instead of an hour before they hit. A wind "profiler" constantly takes the pulse of winds with two overlapping antennas. The two readings meet in a computer to determine the speed and direction of masses of air. Eventually, the PROFS system will narrow forecasts down to a five-square-mile section. Home computers will be wired to a forecasting system

For many areas of the world, the polar-orbiting satellites' Automatic Picture Transmissions (APT) system is the only source of weather data. There are more than 900 APT stations located in 122 countries, including the Communist block of nations. Shown here is a direct readout image received in Italy. (NOAA)

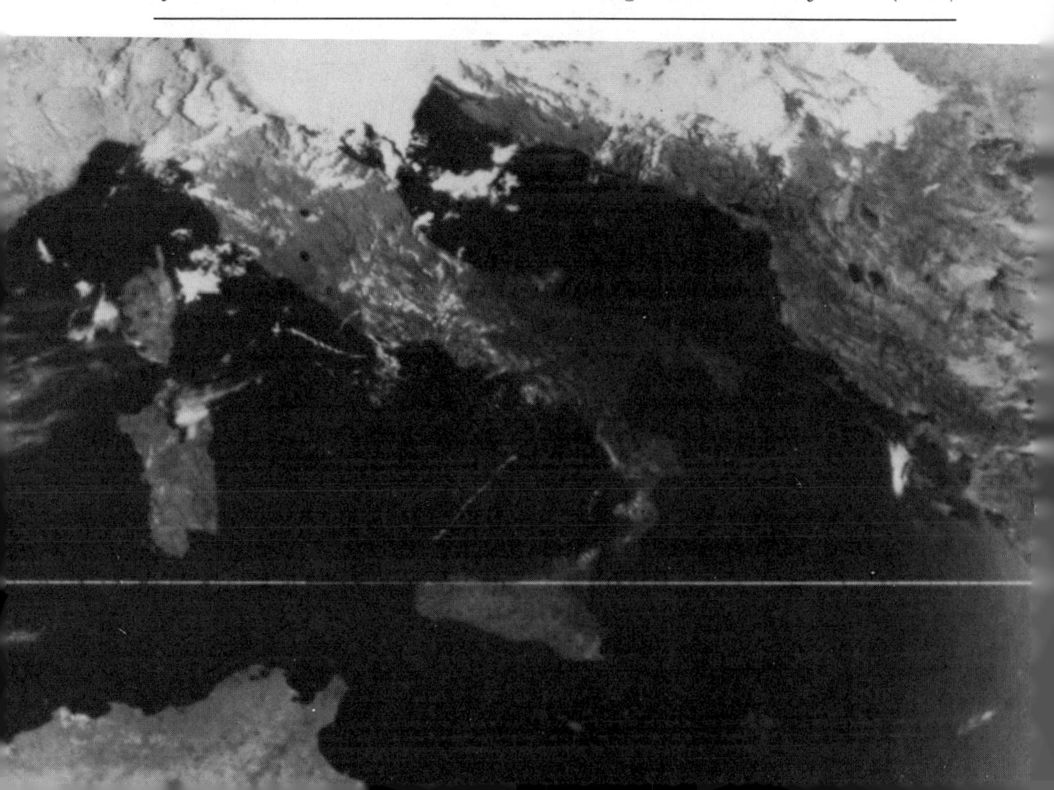

like PROFS so that a person can ask, "What's the temperature going to be in my backyard tomorrow for our barbecue?" The answer will be exact.

With sudden and devastating force, thousands of tornadoes rip mainly across the Midwest each year, usually without warning. They are hard to predict, and often the first time a weather station knows of one is after it's been seen and heard. A new kind of radar called the Doppler radar system will be installed beginning in 1988 to detect tornadoes. On conventional radar screens, the different colors show how much rain or other precipitation is falling at a specific place. The new Doppler radars show how fast the storm is moving toward or away from the radar. It works like the traffic radar used by police. The radar beam bounces off a moving car bumper or a moving raindrop, and its frequency either increases or decreases depending on the speed and whether it is moving toward or away from the radar receiver. The experimental Doppler radars have predicted 70 percent of the tornadoes in an area, giving forecasters a twenty-minute headstart to warn residents. Even a few minutes' advance notice can save lives.

A world-wide weather watch will be more effective from the astronaut-tended Polar Platform planned as part of NOAA's space station program through the year 2000. Two platforms will take measurements of the atmosphere and oceans four times a day, twice during the daylight hours and twice at night as they crisscross over the equator and poles. The platforms will carry research payloads for simultaneous study in many fields.

NOAA intends the Polar Platform to be a truly one-world station funded by most of the big nations including Japan, Canada, the Federal Republic of Germany, Italy, the United Kingdom, and the United States. Russia now cooperates in the satellite search and rescue missions; it's possible it might also take part in the Polar Platform one day.

NOAA's report on the space station plan closes with this statement, "The peaceful daily observation of the earth's atmosphere, oceans, and land from a Polar Platform—provided through international cooperation and tended by a multinational crew of astronauts—is one of the dreams of the Space Age."

Faster, more accurate weather forecasts will be only one advantage of the new space technology. The view from space has changed the way we see our planet. No longer can we think of it as separate compartments of land, water, air, and life. As we discover more about the thin blanket of air that separates us from space, we will be able to answer some of the most puzzling questions about the long-term weather patterns that make up our climate.

3

THE ICE AGE PREDICTION

On Friday, January 28, 1977, the people of Buffalo, New York, were ready to believe that the Little Ice Age had returned.

It had been a cold, snowy winter, but the towns and cities bordering the eastern end of Lake Erie are accustomed to that. Fleets of snow-removal equipment are dispatched almost as soon as the first snowflake falls. No one panics.

That Friday was different. Snow fell steadily all morning. Newscasters warned people to leave a little earlier for work, to be careful shoveling, and to take extra precautions driving, but there was an added urgency to the warning. Gusts of wind up to 70 miles an hour were sweeping 10,000 square miles of snow from the frozen surface of Lake Erie and dumping it on western New York and southern Canada. By nightfall, thousands of people were buried in cars, stranded in schools, factories, or offices, and some, more fortunate, in restaurants and homes of helpful strangers. Emergency conditions existed for days.

Snowdrifts higher than the barns isolated cattle until farmers could tunnel through to them. At the Buffalo Zoo, moats filled quickly with snow, making an easy escape route for animals. Three Scandinavian reindeer walked over the top of one enormous drift, to be captured after a wild snowmobile chase on a college campus

several miles from the zoo. Some of the bison, antelope, and other hoofed stock died in spite of the National Guard's efforts to deliver hay and grain to them. Only the Siberian tigers and polar bears seemed to enjoy the snow as much as the reindeer.

On Saturday, February 5, 1977, President Jimmy Carter declared nine counties a major disaster area, the first such declaration ever made for a snow emergency. It was a time of tragedy and heroism. The National Guard sent men and equipment to help dig out the city, but they ran out of places to pile the snow. Finally, they shipped it out of town. Railroad cars loaded with snow chugged southward, trailing water along the route as the cargo melted. Snowdrifts, black from accumulated dirt and debris, stood far into May as reminders of that strange winter.

◀ *An aerial view of Buffalo, New York, during the blizzard of 1977*
(U.S. ARMY CORPS OF ENGINEERS, BUFFALO DISTRICT)

Cars were buried under snow for weeks following the blizzard in Buffalo, New York, in 1977. (BARBI LARE)

When the blizzard of January 1985 swept through Buffalo, closing schools and businesses for a week, it was not as devastating as the 1977 event because plans had been made for such an emergency. A driving ban in force for five days allowed plows to clear the 700 miles of streets within city limits. A National Guard unit, on its way to an Arctic training session, changed its plans and took its equipment to Buffalo to haul away 400,000 tons of snow.

If Buffalo was prepared, cities to the south were not. The governor of Florida declared an emergency when citrus crops were destroyed in the worst freeze of the century. The jet stream swept south with a surge of frigid Canadian air that broke or tied twenty-seven low-temperature records from Tennessee to Texas. Snow fell in Sarasota and St. Petersburg on Florida's Gulf Coast. The inauguration parade for President Ronald Reagan was canceled to the disappointment of more than 11,500 participants. Europe shared the cold. Children from Paris to the Riviera played in snow for the first time.

Many explanations were offered for the extreme weather patterns the year of the 1977 blizzard, including the far-reaching effects of a warm ocean current off the coast of South America, with widely different speculations on 1985's cold start.

Long-term weather is relatively easy to predict. It's safe to say that January will be colder than June, or that the next ice age will last 100,000 years. The hardest weather to predict is a week away. Short-term weather is elusive, darting away on cold fronts, evading the best of forecasters.

The weather we have taken for granted as "normal" in recent years has actually been unusual. As climatologists read the evidence, the world is going into an ice age after a brief interglacial period, which has been a kind of calm before the storm. This ice age is not on our doorstep. We are not about to be buried under glaciers. And while the trend of climate moves toward a colder world in the next two hundred years or so, there are many events that change our seasonal weather.

Daily, even hourly, weather reports are so routine that it's difficult to imagine life without them. At the touch of a button, we can find out the temperature in Hong Kong, Paris, or Wichita. We know whether to prepare for frost, tornadoes, or a heat wave. Will it be a white Christmas, a clear day for a shuttle lift-off, or a good summer for growing corn? Ask the weatherman.

Meteorology is not an ancient occupation. We didn't even have reliable thermometers until 1714 when Gabriel Fahrenheit figured out how to clean mercury so it wouldn't stick to the sides of the narrow thermometer tube. He devised a scale for measurements still in use today.

The seventeenth century marked one of mankind's first big scientific spurts. In one decade, six important scientific instruments were invented: the telescope, microscope, air pump, pendulum clock, thermometer, and barometer. None of these instruments was commonly used, nor were any of them extremely accurate. Early weather records provide a general picture but not the detailed, accurate data we expect today. For one thing, the calendars were not standardized, and some reports were dated as the year of the reign of some prince or princess of a territory that may have become part of a larger country, so that it is difficult to know where and when it all happened.

Hubert H. Lamb started the Climatic Research Unit at the University of East Anglia in England to begin what he calls a thousand-year program to map the weather across Europe using just historical records. His book, *Climate: Present, Past and Future*, describes the weather long before weather records were kept, taken from old ships' logs, letters from soldiers, records of farmers and tradesmen, old journals, church records, and especially ledgers from tax collectors. The kind of journal entry he finds includes a notation that Bastille Day was a "good day for outdoor activity."

Greenland was really green in the days of Eric the Red and other adventurous Norsemen. The great Icelandic sagas handed down from one generation to the next are stories of settlers in Iceland and Greenland in the eighth and ninth centuries after the glaciers from the great Ice Age had retreated and the open coasts were lush with forests. There is evidence that this warm climate in the Northern Hemisphere lasted well into the 1300s. For the first time in history, the North Sea and the North Atlantic were free of drift ice and safe for travel. The Norsemen set out in their elegant open ships to look for new sea routes and new lands.

The Vikings followed. They were the Scandinavian raiders who plundered ships and settlements all along the western coast of Iceland and the southwestern coast of Greenland, where deep fjiords provided sheltered harbors from winter storms. Two centuries of exceptional warmth in the north helped the Vikings take over Iceland and Greenland and sail on to Newfoundland in North America.

By the tenth century, the colony founded by Eric the Red had grown to 280 farms with more than 3,000 settlers. They raised food for themselves with enough left over to sell on regular trade routes to the European markets. Unusually heavy rainfalls were recorded during this warm period in England and Europe. During many springs farmers could not get to their fields to plant because of the thick mud. Often torrential rains in the fall damaged crops or made harvesting difficult.

Winters, too, were damp, but extremely mild, perfect conditions for a blight of fungus to thrive on fields of rye. Whole villages were afflicted by the dreaded St. Anthony's fire, a disease caused by the fungus. Even a few kernels of this blighted rye, made into flour, baked into bread, and eaten, was enough to cause hallucinations and madness. People, hallucinating, danced in the streets, and because no one knew the cause, it was thought to be the work of an evil spirit. Even dogs and cats died after eating the contaminated bread. The disease raced across Europe, and in 1096 the Order of Hospital Brothers of Saint Anthony was founded to care for the ill and dying. The disease became known as St. Anthony's fire because, as it progressed, the victim's fingers and toes turned black and fell off as though charred by flames. Today we know the disease is caused by a fungus alled *Claviceps purpea,* which produces a hallucinogenic drug similar to LSD.

But all was not hardship during the warm centuries. Vineyards flourished in parts of England where grapes had never grown before nor since. Wheat grew in Iceland and in sections of northern Europe where it cannot be grown now.

North America felt the same warming trend. Prior to the 1200s, the Native Americans of the Midwest were farmers living in small villages and raising corn and other crops. Just east of St. Louis, archaeologists have uncovered the remains of a huge village that once housed perhaps 40,000 people. What happened to the Indian farmer, the colonists of Greenland, and the vineyards of England?

Climatologists call those years the Little Ice Age. The climate cooled toward the end of the thirteenth century, a time when "some civilizations began to stumble," says climatologist Stephen Schneider in his book *The Coevolution of Climate and Life*, "because in Europe as well as many other parts of the world the extent of ice and snow on land and sea seems to have been as great or even greater than at any time since the end of the Pleistocene." From the middle of the 1300s through the 1700s, drift ice increased and blocked the trade routes between Iceland, Greenland, and Europe. Greenland, especially, became isolated, and there is no record that it opened to ships again until a brief warming period in the 1500s.

When ships did return, it was to empty harbors. People had disappeared. Villages and farms were deserted, decayed, and covered with ice. Whatever survivors there were had apparently taken up the life of the nomadic Eskimos and scattered to other regions.

When Danish archaeologist Paul Norland explored the abandoned Greenland villages, he excavated gravesites. Graves are never dug below permafrost, which is the level of ground that is always frozen. Norland found the oldest graves deep in the soil at the old level of permafrost. Skeletons were perfectly preserved along with clothing and wooden objects that had remained unchanged for five hundred years. He found the more recent graves shallower, and the last graves barely beneath the surface, showing that the ground had frozen quickly. It was stark evidence of the change in climate.

The settlements of Iceland fared better than those of Greenland, where glaciers moved southward, covering many farms. With shorter growing seasons, few crops survived, and the remaining farmers settled along the coast, where they could fish. But even the hardy cod were less abundant in the colder Arctic seas.

The Little Ice Age was a time of extremely variable, unstable weather for several hundred years. Along with severe storms, late springs, early falls, and cool summers, it brought drought to the American Midwest.

Archaeologists Reid A. Bryson and David Baerreis excavated several villages of the Mill Creek Indians who had been farmers. They found great numbers of deer and elk bones, evidence that the Mill Creek people had relied on animals for food when crops failed. In more recent sites, they found fewer deer and elk bones, but more bison skeletons. Deer and elk are browsers that feed on shrubs and trees. Bison are grazers, feeding on grass. It was an indication of sparse rainfall because trees and shrubs need large amounts of water, while grasses can grow in drier conditions as they do now in the Midwest. When the wagon trains of early European settlers moved west, they met Native Americans who had become hunters because they had to abandon their farms to find food.

The general cooling in Europe and North America went on through the 1700s. Tax records from a parish in Chamonix in the French Alps show that a large parcel of land had to be removed from the tax roles because an advancing glacier had wiped out an entire village along with twelve houses from the neighboring town. A notation made in 1640 said that the glacier was advancing more than the distance of a musket shot each day.

The recovery from the Little Ice Age was erratic. The winter of 1708 was so cold that parts of the Baltic Sea froze, and in 1716 people skated through London on the frozen Thames River. Through those same years the summers went from one extreme to another. Some of England's wettest summers hit in the mid-1700s while France experienced severe droughts. With scrawny harvests, food prices soared, and a typical French worker had to spend up to three-quarters of his income on bread alone. In that way the changing climate added fuel to the fires of the French Revolution.

American colonists wrote home about severe weather, too. Governor John Winthrop's journal of 1637 says, "This is a very hard winter." That year Boston harbor froze, and snow fell the first week of May. When the Hudson River iced over in 1705, 132 sailors died in their ship, which was grounded off Sandy Hook. The Hudson again froze in 1790, and even the mighty Mississippi was filled with ice at New Orleans in 1784.

Because the crazy weather conditions concerned not only farmers but scientists as well, several countries began collecting weather statistics. The Swiss, French, and English were among the first. Although no official records were kept in the New World, the Reverend John Campanius Holm faithfully kept daily records from his home along the Delaware River in the mid-1600s. Today the National Weather Service honors a volunteer weather observer each year with the John Campanius Holm award.

When it became important to know weather conditions during the War of 1812, Army surgeons were charged with logging daily wind directions, velocity, temperatures, and precipitation. Later that job fell to the Signal Corps, until the United States Weather Bureau was established in 1890 as part of the Department of Agriculture. In 1940 it became the responsibility of the Department of Commerce, and now the National Weather Service is part of NOAA.

The end of the 1800s seemed to be the end of wild weather, and for the next fifty or sixty years things evened off into milder winters, fewer storms, and more pleasant summers with fairly even rainfall. European glaciers retreated from valleys, melting back into the mountains, and farms and villages were uncovered to prosper again. In Africa and Asia, the monsoons became more reliable and even brought rain to usually arid lands. It was a return to what we like to think of as "normal" weather . . . mild, pleasant, comfortable picnic weather with long, profitable growing seasons.

In scientific terms, "normal" weather is only the average of the last thirty years. Normal temperature for October in a specific place in 1985 is the average of temperatures that month since 1955.

The mild, steady climate of the first half of the twentieth century was actually the least normal weather pattern in several hundred or even thousand years. At least nine times in the past million years there have been massive ice sheets covering most of North America and Europe. Ice seems to be the Northern Hemisphere's normal condition. Each cycle of ice build-up and retreat lasted tens of thousands of years, although the climate shifts that set off those changes may have taken only a few hundred years. There have been at least four epochs in the past billion years, which is only a quarter of the earth's age, when ice covered major portions of the planet. We are still in the midst of the fourth one.

We're back, since the 1970s and 1980s, to erratic weather like the blizzard that buried Buffalo, and the killing droughts in Africa, Siberia, and our own Midwest. There seems to be no rhyme or reason to the weather. While the blizzard of '77 raged in the east, in Alaska the polar bears didn't even bother to hibernate because the winter there was so mild.

Such "crazy" weather is typical of an interglacial period with short bursts of cold weather like that during the thirteenth and fourteenth centuries, the time that has been labeled the Little Ice Age. Interglacials, the time between ice ages, last about 10,000 years. We're coming to the end of one of those periods now. Sometime within a hundred centuries, we'll be deep into this planet's normal condition, an ice age. The gaps in the ice ages, the interglacials, are favorable to humans, and it's been in these gaps that civilizations began to thrive. Rather than asking, "What causes ice ages?" maybe we should be asking, "What causes the favorable short warm periods?"

Many combinations of events change climate, but the most underestimated are the changes caused by man himself. Some climatologists say there is a new factor, a "wild card" never there before because, for the first time in history, human activities are beginning to affect weather and climate with a force equal to that of natural events.

4 THE GREENHOUSE PREDICTION

"I've got salt in my well now. Don't tell me the sea's not creeping in."

"Aw, you environmental nuts. You figure we'll fall for the scare tactics. You probably want to buy up the beach property."

"There's not going to be a beach!"

The town clerk pounded his gavel. "Order. This town meeting will come to order or we'll clear the hall."

In the rear of the room one old fisherman, who had lived in this coastal town all his life, stood up. "I read one place," he said, "that we're going back to some kind of ice age. Then you come in here and tell us it's getting hotter 'cause of some kind of greenhouse and the icecap's going to melt. Well, I say there's not a durn thing you can do about it anyway, so why all the hullabaloo?"

All the hullabaloo is about planning ahead.

By geologic time, measured in millions of years, the greenhouse effect may be only a passing phase, but it can change the way we live out our short moment of time on the earth.

The "greenhouse effect" is not an ad agency slogan of the 1980s. The phrase was coined a hundred years ago by the Swedish Nobel Prize scientist Savante Arrhenius, when he described how carbon dioxide formed a "heat trap" in the atmosphere. He predicted that the earth's temperature would rise as the amount of carbon dioxide in the atmosphere increased. And that's exactly what's happening now.

The Environmental Protection Agency and the National Academy of Sciences both issued reports in 1983 warning of severe climate changes from this greenhouse effect. They defined it as the gradual warming of the atmosphere caused by increased carbon dioxide from the burning of fossil fuels. If the amount of carbon dioxide in the atmosphere doubles, which it seems to be doing, the earth's temperature will rise at least 2.7 degrees Fahrenheit (1.5 degrees Celsius), but probably not more than 8.1 degrees Fahrenheit (4.5 degrees Celsius). The World Meteorological Organization pointed out that other gases, which are increasing along with the carbon dioxide, may add enough to that warming to raise the earth's total equilibrium temperature as high as 16.2 degrees Fahrenheit (9 degrees Celsius).

The EPA report, written by Seidel and Keyes, says, "A 3 degree C warming would leave San Francisco as warm as San Diego is today. A 9 degree C warming would raise New York City's average temperature to the current temperature of Daytona Beach, Florida."

How does it happen, and when?

Carbon dioxide is not a dangerous gas, not a threat to human health. It is heavy, colorless, and odorless. Hundreds of billions of tons of it are exchanged each year between plants and animals, between sea and air, in an unending cycle. Our sister planet, Venus, is a superhot desert because of a runaway natural greenhouse effect produced by its thick atmosphere of carbon dioxide. That's unlikely to happen here.

The sun's energy strikes the earth mostly as visible or ultraviolet light. Most of it is absorbed in the highest layers of the atmosphere by molecules of nitrogen oxides, oxygen, and especially by ozone (a three-atom molecule of oxygen). Without this filtering-out process, there would be no life. High doses of ultraviolet could destroy DNA, which is life's foundation.

More of the ultraviolet light is reflected back by cloud cover, and some is absorbed by the earth. As the earth heats up, some

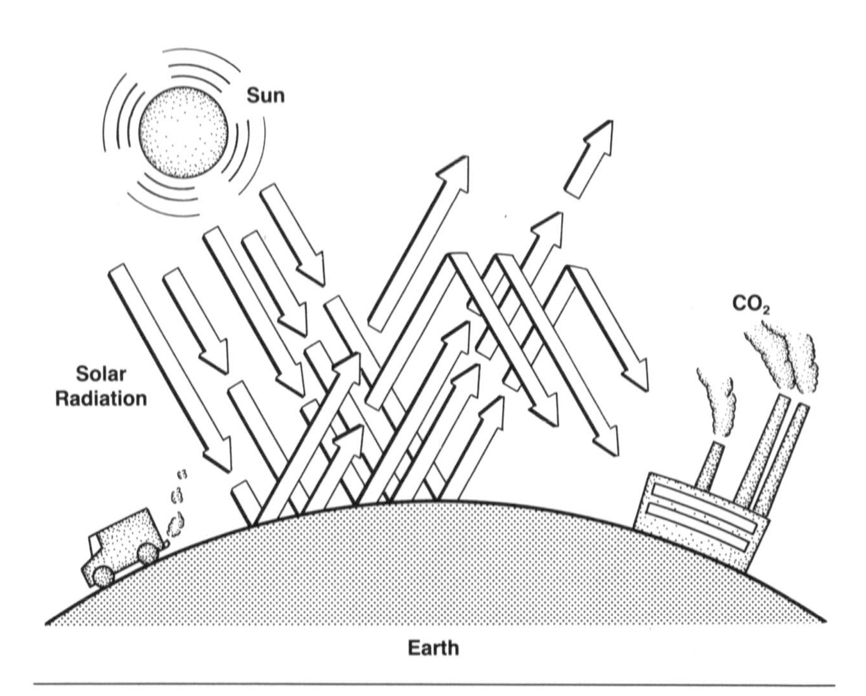

The greenhouse effect warms the earth as we burn more fossil fuels and add more carbon dioxide to the atmosphere. The carbon dioxide traps infrared (heat) rays, and less heat is radiated back into space.

of the sun's energy is radiated back into space, not as visible light energy, but as longer wavelengths of infrared or heat. Carbon dioxide in the atmosphere lets visible light through, but it absorbs the infrared wavelengths. The more carbon dioxide in the air, the more heat is kept in, just like glass traps heat in a greenhouse.

Nobody paid much attention to Arrhenius's theory at the turn of the century because there was little practical reason to do so. Now there is.

The first accurate measurements of carbon dioxide came from the Scripps Institute of Oceanography in 1957 at its atmospheric observatory far from industrial pollution on the volcano of Mauna

Loa in Hawaii. In 1958 carbon dioxide in the air was measured as 315 parts per million. In 1984 it had risen to 340 parts per million. Other research stations studied pockets of air trapped in glaciers, which showed that the concentration of carbon dioxide had been only 265 parts per million in the middle of the nineteenth century. Why the dramatic rise?

We've been burning fossil fuels. Coal, oil, and natural gas are made of carbon. This "old" stored carbon is put back into the atmosphere as carbon dioxide, a by-product of burning, and it's put back into circulation suddenly at a rapid rate. Five-and-a-half-billion tons of carbon are released into the atmosphere each year.

Great reservoirs called "sinks" store the carbon dioxide. The atmosphere, the oceans, and all plants and animals, which together are called the biomass, are sinks for this gas. Oceans can absorb 80 percent of the carbon dioxide from fossil fuels, but we're burning it faster than the oceans seem to dissolve it. Enormous stands of forests absorb it, but we're cutting timber faster than we're planting new forests.

The biggest uncertainty is the role of the oceans. Although the seas may contain sixty times as much carbon as exists in the atmosphere, researchers are just beginning to find out how long it takes huge bodies of water to absorb carbon dioxide or to warm up. The upper layers of water heat quickly and exchange carbon dioxide quickly, but it takes much longer to transfer that heat and gas to the deepest regions of the sea. At the poles, cold water sinks to the ocean bottom and moves toward the equator in a cycle that is estimated to take a thousand years to make a complete circuit.

If we had a twin earth to practice on, it would be easier to find and control the ways in which we change our atmosphere, but the closest we have is a computer model. Climatologists can create a sample world or model on a computer by programming into it

---

Industries add carbon dioxide to the atmosphere.

(FAO PHOTO BY CAMILLO BOSCARDI)

statistics that represent a spinning planet, energy from the sun, the role of the oceans, and the chemistry of the atmosphere. They sprinkle in variations—add carbon dioxide from factories, toss in some volcanic ash and some disappearing forests—and the computer shows what might happen to temperatures, to rainfall, to melting ice. They make mathematical changes to represent the sun's intensity, perhaps, to see what happens to world-wide weather.

These climate modelers are not ordinary household computers, but so enormously sophisticated and expensive that only a few laboratories have access to them. And how do the researchers know if their calculations are right? They compare their simulated climates with CLIMAP to see if they match what actually happened in the past. CLIMAP is a federally funded project, Climate: Long-Range Investigation, Mapping, and Prediction, designed to chart weather patterns of the past in order to learn what may happen in the future. The enormous map drawn by CLIMAP shows the world as it was during an average August in the last glacial advance 18,000 years ago from information gathered from ancient soil layers, ice samples, deep sea-floor samples, tree rings, and many other sources.

With these computer models climatologists can produce what they call "worst-case scenarios," "best-case scenarios," or "business-as-usual scenarios." Science writer John Gribben says in his book *Future Weather and the Greenhouse Effect*, "Only a fool would take the resulting 'guestimates' as gospel truth about what lies ahead; but only a bigger fool would ignore the problem entirely in the hope that it might go away of its own accord."

If, as the EPA report says, global temperatures rise 8 degrees Fahrenheit (4.4 degrees Celsius) by the end of the next century, it will be warmer than at any time since the last ice age. It will change the annual rainfall, raise the level of the oceans, either swell or dry up rivers, disrupt agriculture, and change nations.

After reading the EPA report, a reviewer for the National Academy of Science suggested that the world may get into trouble in ways we can't even imagine yet.

Alpine glaciers and the polar land ice could melt by the year 2000, and seas would rise two feet, flooding low-lying communities.

Drought would strike some now-fertile farm regions, and some deserts would get enough rain to raise crops.

A four-degree heating, says the NAS report, would dry up the water in America's western rivers because runoff water would evaporate before it worked its way to the rivers.

For every region put in jeopardy by the warming, another might prosper. If the Arctic ice packs melt, it could open the fabled Northwest Passage, and oil and gas exploration could move into the Arctic Ocean more easily. The Soviet Union stands to benefit because it is likely that frigid parts of its country could become farmable, while China and India are likely to suffer more drought. The high-yield grain-producing farms are likely to move to the northern plains of Canada, while Kansas, Missouri, and other states in today's grain belt could turn into another dust bowl.

The big industrial nations are most responsible for the carbon-dioxide build-up, and they are the ones who must cooperate to work out a solution. Based on a "business-as-usual scenario," meaning we wouldn't change anything, today's biggest users of fossil fuels are in danger of becoming the poorest nations.

Veerabhadran Ramanathan, a climate modeler with a team at the National Center for Atmospheric Research, warns that, by the year 2030, the combined effects of trace gases such as nitrous oxide, methane, and the chlorofluorocarbons (CFC-11 and CFC-12) will magnify the greenhouse effect from carbon dioxide by up to 30 percent. He compares that to making the sun shine brighter by 1¼ percent for the next fifty years.

The thirty trace gases they studied add to the global heating because molecule for molecule they are much more effective at trapping infrared energy. The chlorofluorocarbons, which are by-products of aluminum manufacture, hang around the atmosphere for hundreds of years, long after carbon dioxide and other gases have broken apart. As a result, even small amounts are dangerous as they build up over time.

In the United States, aerosols with CFCs were banned in 1978, but other countries continue to use them. Other nations also use CFCs in refrigeration and air conditioning. The trace-gas study also predicts that changes in the ozone will add greatly to the warming of the earth. Ozone in the upper stratosphere will drop, allowing more sunlight to reach the earth, while ozone in the lower altitudes is increasing, which will add to the greenhouse effect.

Pollution from industry, cars, and jets pours sulfur dioxide, water vapor, and oxides of nitrogen into the air. Tons of these pollutants wash out in acid rain and snow, which poisons lakes, streams, forests, and soil. They etch their way through buildings and kill fish and other wildlife.

Methane gas is one of the trace gases that adds to the greenhouse effect as it absorbs infrared light rays. Dr. F. Sherwood Rowland, who was the leader for a team of researchers at the University of California, Irvine, for a seven-year world-wide study, believes that concentrations of methane are rising because of a growing need to feed the earth's rapidly increasing population. The number of cattle has gone up 50 percent, and methane gas is released into the air by cows during the digestion of cellulose from the grains they eat.

Rice production has increased greatly in the last few years, and the flooded rice patties produce methane known as "swamp gas," that pungent odor common in marshes and bogs.

A molecule of methane is twenty times more effective at trapping heat than a molecule of carbon dioxide,and the earth's temperature depends on the number of atmospheric molecules such as methane and carbon dioxide that absorb the heat it radiates. With too few such molecules, the earth would cool, and too many would heat it.

Our energy crisis is partly caused by the ways in which we handle our fossil fuels and the rate at which we burn them. The EPA report said that even if governments put a 300 percent tax on fossil fuels, it would delay a 2-degree Celsius warming only until the year 2045. And even if we put a total ban on the use of shale oil and all synthetic fuels, it would delay this 2-degree Celsius warming only five years.

In other words, the warming is on its way, and short of stopping the use of any fossil fuels, there's not much we can do about it now except get ready.

Like the people at the hypothetical town meeting that opened this chapter, concerned citizens are discussing the problem. Cape Cod, Massachusetts, is only a few meters above sea level, and land developers are concerned about water supplies. Fresh water hangs beneath the cape in a fragile layer that could be easily contaminated by salt water moving inland. Planners must also decide whether to build barriers to hold back the sea or to move structures inland.

It may not be possible to stop the flooding of coastal cities, but engineers might design dikes that could be installed to cut the damage. Catch basins and reservoirs could be built in areas where droughts are expected, and agricultural engineers could breed drought-resistant crops. It usually takes ten years to build power plants, and now there is time to plan them for safe sites away from eroding coastlines.

We can begin to develop active solar power and other systems not dependent on fossil fuels. Some climatologists suggest that a good start might be made by requiring government buildings to operate on solar energy.

Scrubbers and other pollution-control equipment on factories will help some, but such equipment is extravagantly expensive and still won't stop the warming trend.

Huge stands of trees can act as sinks to absorb great quantities of carbon dioxide. We know that sycamore trees, for example, can absorb an average of 750 tons of carbon annually for each square kilometer of land. But to offset fifty years of carbon-dioxide emissions at the rate they are pouring into the atmosphere, we'd need to plant 6.7 million square kilometers of trees, which would require an area roughly the size of Europe. About half the land on this planet is already covered with trees or used for agriculture or cities, and much of the remaining land is sand, rock, or ice. Even if a place could be found to plant enough trees, it would need enormous amounts of fertilizer, as much 17 million tons of phosphorus and 10 million tons of potassium (potash), which is about a third of the fertilizer produced now. And all of it is needed on farms.

Several energy sources are being considered to replace fossil fuels. Nuclear fusion is a process that extracts "heavy" hydrogen from water and combines it with two hydrogen atoms to form helium.

Hydrogen can be generated by "splitting water" through the use of solar energy directly or electricity (electrolysis). On a very small scale, splitting water is one of the experiments commonly done in high school science classes.

Solar satellites can collect solar energy in space and transmit it to earth by microwaves.

Green plants might even be tapped for energy some day through bioengineering. If a super species of high hydrogen-producing plant could be genetically engineered, it might add to our energy supplies.

The EPA closes its report by admitting we can only guess at the results of global warming, but they say, "Innovative thinking and strategy-building are sorely needed. A 2 degree C increase in temperature by (or perhaps well before) the middle of the next century leaves us only a few decades to plan for and cope with a change in habitability in many geographic regions."

In other words, we can't ignore what's happening; we must plan for changes. It also means we must keep up the research because the greenhouse effect is only one small part of the earth's changing climate.

5

THE OCEAN CONNECTION

At midnight the wake of the *Wecoma* foamed and sparkled in the moonlight. Several biologists and technicians from Duke University's Marine Laboratory, Oregon State University, Peru's Instituto Nacional de Pesca, and Ecuador's Instituto Oceanographico Casilla stood watch on the ship's fantail enjoying the cool equatorial breeze. For seventeen days in April 1985, small teams rotated duty around the clock. They were collecting and processing seawater samples for a study of equatorial currents.

When the ship reached the degree of latitude and longitude plotted for one of the thirty-six stations in the research plan, it stopped. A cluster of twelve cylinders mounted on a frame called a CTD (for measuring conductivity, temperature, and depth) was swung over the starboard side and lowered into the sea. Computers in the ship's lab electronically tripped the numbered cylinders to open and fill at specified depths. When the CTD was hauled aboard, the technicians swarmed around it, filling small plastic bottles with seawater they would filter and analyze in the ship's lab.

The *Wecoma*, operated by Oregon State University, was following a course around the Galapagos Islands, which lie on the equator 650 miles west of Ecuador. It was not the fascination of these unique islands, which Darwin called a "living laboratory of evolution," that

◀

Pablo Intriago, a student at the Instituto Nacional de Pesca in Guayaquil, Ecuador, pulls in a bongo net that was trailed beyond the research ship to collect samples of plant and animal life at various depths. (MARGERY FACKLAM)

Scientists on the research vessel Wecoma *pull in the CTD equipment, which collects samples for studying conductivity, temperature, and depth of the water. Each tube is triggered by a computer in the ship's lab to open at different depths and fill with water.* (MARGERY FACKLAM)

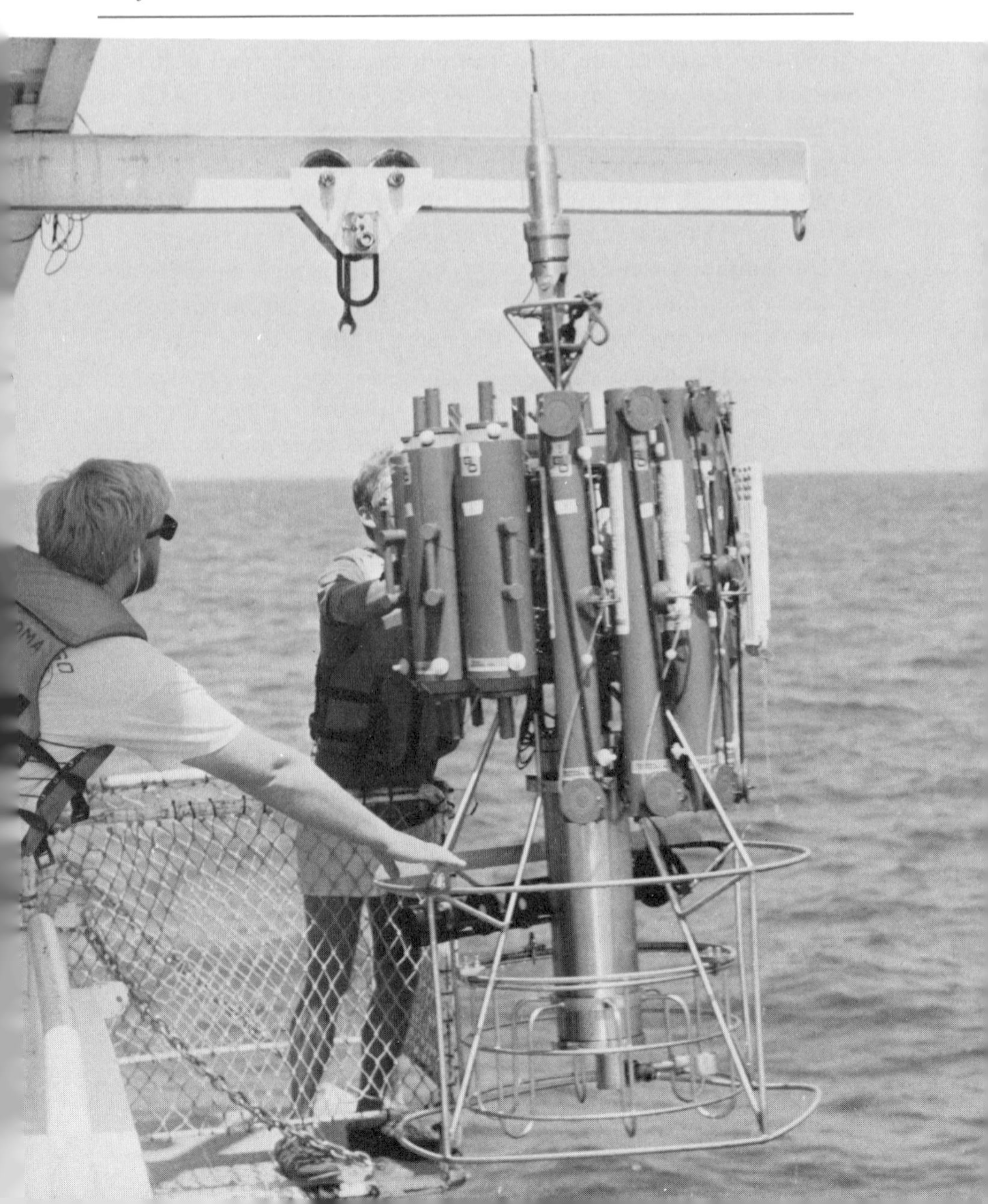

lured the scientists to this part of the Pacific, but an ocean current called El Niño.

El Niño, the child; it's a strangely misleading name for a phenomenon that triggered weather changes around the world in 1982–83 like a chain of firecrackers. For centuries it was known simply as a warm current off the coast of Peru that periodically mixed with the cold Humboldt Current, bringing with it drenching rains to an otherwise arid coast. When fishermen saw the banana leaves, branches from palm trees, and other flotsam floating southward late in December, they knew it signaled the end of fishing season. They called it Corriente del Niño, or Current of the Christ Child because it usually occurred at the Christmas season. Every few years the current is stronger and warmer than usual, but in 1982 it was "in many ways one of the most unusual events of the century." That was the opinion of Dr. Eugene M. Rasmusson, a NOAA authority on El Niño, who said, "There've been eight significant El Niños since World War II. They occur on average of every four or five years, but irregularly—they can be two years apart, or as many as ten."

The El Niño of the eighties caused disastrous events in places far from its origin: drought followed by devastating dust storms and fierce bushfires swept Australia, India, and Africa; rare hurricanes and cyclones peppered the Pacific; across the United States, 1983 had the rainiest spring ever recorded. California had triple its normal rainfall, and floods and mud slides destroyed property along the West Coast. The coast of Peru and Ecuador is virtually desert, and four or five inches of rain is normal. El Niño caused a total of twelve feet of rain in 1983, wiping out roads and bridges, washing out adobe houses, and killing people in flash floods and mud slides. By the end of 1983, El Niño had taken 1,100 lives and had cost more than 8.7 billion dollars in damages to land, crops, livestock, and marine life.

In the wake of El Niño, heavy rains in Utah caused devastating mud slides that destroyed many roads and homes. (© 1983 DAVID CORNWALL/BLACK STAR)

Other, less devastating situations ranging from expensive anchovies for American pizzas to increased shark attacks on swimmers also resulted from El Niño's weather changes. In Montana, the number of people bitten by rattlesnakes increased as the hot, dry conditions caused mice to move to more populated areas to find food, and the rattlers followed. Bubonic plague cases in New Mexico increased because the unusually cool, wet spring allowed both the rodents and the fleas they carried to thrive. Fleas carry plague bacilli.

Dozens of commercial fishermen from southern California north to Washington suffered great losses because El Niño changed the sea temperature. The plankton the fish feed on grow best in cool water, and when there is no plankton, fish either die or move to different feeding grounds. The same warm currents lured sharks closer to shore in search of food and so increased the danger to swimmers.

Seventy percent of the earth's surface is water. Science-fiction writer Arthur C. Clark has been known to say, "How inappropriate to call this planet Earth, when clearly it is Ocean." Only a thin sheet of water separates the atmosphere from the ocean depths. The difference in temperature between that sea surface and the air above it is what causes heat to flow and water vapor to be exchanged. Dr. Richard T. Barber, professor of marine biology at the Duke University Marine Laboratory, says, "The ocean is clearly driving the atmosphere."

We know now that El Niño is more than a warm current coming from the north and mixing with the Humboldt Current. It is a complex interaction between ocean and atmosphere, a chain reaction with far-reaching effects on world-wide weather and the economy. The phenomenon is technically known as ENSO, El Niño-Southern Oscillation, but everyone calls it just El Niño.

Years after the devastation of the 1982 El Niño, Peruvians still talk about the terrifying thunder and lightning, which they had never seen before, that accompanied the torrential rains. In Puerto Ayora on Santa Cruz, the only town in the Galapagos Islands, residents remember 1983 as the year the rainfall records went off the graph. In May rain fell steadily for twenty-six days for a record 27.5 inches, where there might not be 5 inches in a year. Streams appeared where water had never flowed before, cascading over volcanic cliffs. One naturalist-guide was only partly kidding when he said that there are probably fewer people who have seen a waterfall in the Galapagos than have climbed Mount Everest. They are normally an arid group of islands. Giant cactuses and the ghostly white palo santo trees grow alongside scrubby shrubs with leaves the size of a quarter. During that year of rain, leaves from those same bushes grew as big as saucers.

Researchers on the 1985 *Wecoma* cruise were doing follow-up studies on the 1982 El Niño and looking for signs of the next one. It was on the 1982 cruise of the research vessel *Conrad* that Dr. Barber had happened upon the beginning of the notorious El Niño, a coincidence that gave scientists a chance to track its growth for the first time.

The *Conrad*'s crew had sailed from Hawaii toward the equator expecting to feast on fresh fish every day. They looked forward to sleeping on deck in the pleasant breeze of the trade winds. Some of the graduate students who had never been to sea hoped to watch humpback whales surface and blow, and pods of dolphins leap in the ship's wake.

But the ocean was vacant. The cooks caught not a single fish as they trolled off the stern. Not one whale or dolphin was sighted, and as the *Conrad* neared the equator, the weather got muggier and hotter. Sea-surface temperatures soared. Nights were so uncomfortable that no one could stay on deck more than fifteen minutes before they were soaked in sweat.

When one of the *Conrad*'s engines quit, they expected to run behind schedule. Instead, the ship was pushed ahead on an equatorial current moving east at one or two miles an hour. But the current was going the wrong way. It was a complete reversal of the normal current that should have been flowing westward with the trade winds.

Back at the National Oceanic and Atmospheric Administration Climate Analysis Center, computers were rejecting the measurements of wind speeds, current speeds, and temperatures that were coming in from drifting buoys, satellites, and ships at sea. The computers had been programmed to reject any measurement that was too far from normal because it might be caused by a defective instrument. But when the same kind of information kept coming in from the *Conrad* and other research ships, NOAA scientists reprogrammed the computers. They knew then they were seeing the start of a major warm episode called El Niño.

Like the ripple that spreads across a pond when you toss a stone into it, weather changes ripple around the world triggered by disturbances in the atmosphere, in the biosphere (wherever plants and animals live), in the cryosphere (ice and snowfields), and in the oceans. Meteorologists call these long-distance links "teleconnections." The most recent shift in El Niño gave scientists a chance to study these teleconnections. The crew on the *Conrad* research ship had stumbled into what they have called "an enormous living laboratory."

One of the first to suspect an atmospheric-oceanic connection to climate was Gilbert "Boomerang" Walker, a British mathematician, who was posted to India in 1904 as Director General of Observatories just as the country was recovering from the devastating famine of 1899–1900.

Monsoons are fickle, and they make life in India something of a lottery. If the rains expected between June and September don't come and there is drought, or if rains are more torrential than usual, crops fail and widespread famine and starvation follow.

Walker, whose nickname came from his interest in boomerangs and other primitive weapons, decided to find a way of forecasting the yearly changes in monsoons. He was convinced they were in some way tied to global weather. Using statistics from weather stations all over the world, Walker set a small army of clerks to computing (without computers) relationships between air pressure, rainfall, and temperatures. He even had them correlate such seemingly unrelated data as the appearance of sunspots and the extent of the annual flooding of the Nile.

Twenty-four years later, Walker came up with a theory of atmospheric oscillations, or regions in which air pressure changes. He noticed that when pressure is high in the Pacific Ocean, it tends to be low in the Indian Ocean. He called that the Southern Oscillation. Not only did he see the relationship between oscillations of air pressure in the eastern and western Pacific and the monsoons in India, but also to rainfall in Africa. And during these oscillations, he noted, the temperatures in western Canada were above normal. Walker was convinced that all these events were part of the same phenomenon.

For the most part, meteorologists paid no attention to Walker's writings because they were skeptical of theories that gave a simple, single explanation for world-wide weather patterns.

Then in 1960, Jacob Bjerknes, a meteorologist at the University of California at Los Angeles, found the link between the sea surface and the air pressure that caused winds and water to change direction. He named it the Walker Connection in honor of "Boomerang" Walker who died that year at the age of ninety.

An oceanographer, Klaus Wyrtki at the University of Hawaii, completed the picture of the ENSO phenomenon when he described how the westerly trade winds normally act as a dam that tends to hold warm water in the western Pacific. When air pressure changes and winds abate, this dam breaks, allowing the warm water to move toward the east.

The "wrong way" ocean current that pushed the *Conrad* was accompanied by changes in sea-surface temperatures. During that 1982–83 El Niño, the worst since 1891, a warm tongue of water stretched 8,000 miles along the equator, heating the water as much as 14 degrees Fahrenheit (minus 10 degrees Celsius) above normal. It raised the sea level eight inches and caused high tides, giant waves, and torrential rains northward along the coast of Mexico, the United States, and Canada. These natural disasters ravaged people and property, but they also altered the life cycles of many species and changed many habitats.

Fishing is good along the Peruvian coast because cold water from the ocean bottom wells up, bringing with it rich nutrients fish feed upon. From the 1950s to 1970s the fishing fleets of Peru hauled in

Off the coast of Peru in the nutrient-rich waters of the Humboldt Current, fishermen gathered in huge harvests of anchoveta (a small member of the anchovy family) before the years of overfishing and before the warm waters of El Niño moved in. (FAO PHOTO)

14 million tons of anchovies each year, which is a fifth of the world's fish catch. During an El Niño event, however, microscopic plankton do not thrive in the warm slab of water covering the surface.

When the 1972 El Niño warmed the water, the tiny anchovies swarmed close to the coast of Peru and Ecuador, and the commercial fishermen harvested 180,000 tons a day. In the rush to cash in on this gold mine of fish, they almost wiped out a species. According to Peruvian conservationists, their anchovies can be listed as endangered. Without the anchovies, sardines moved in to fill that niche in the food chain, and they began to thrive. The fishing industry turned to sardines, but when the 1982 El Niño warmed the waters, the food chain collapsed once again. Sardines disappeared, many boats rusted in the harbors, and the fish-meal factories closed.

The disappearance of anchovies and other fish in Peruvian waters was only one example of biological upsets on land and sea. Ralph and Elizabeth Anne Schreiber, ornithologists at the Natural History Museum of Los Angeles County, had been studying birds on Christmas Island in the middle of the Pacific for fifteen years. Where they had counted 8,000 greater frigate birds early in the summer of 1982, they found only 100 in November that year. Among all the eighteen species on the island, they found thousands of dead and dying nestlings abandoned by parents who had gone to look for fish for themselves. The Schreibers concluded that in a natural disaster, such as El Niño, animals abandon their young to concentrate on their own survival or cut out reproduction temporarily. Many of sea lions of the Galapagos Islands left their pups to search for food for themselves, which is the kind of behavior that allows a species to survive a disaster to breed again when conditions are more favorable.

Not all species suffered. Some not only survived but became more productive. Although the populations of twelve species of fish went down to almost nothing, nineteen other species expanded their territories and flourished. Tropical tuna and bonita, which are predatory fish found thirty miles offshore, were carried by El Niño close to the South American coast, where they found new prey, and fishermen found new game. The shrimp population exploded close to shore, and a local plant, which processed five tons a week before the 1982 El Niño, turned out seventy-five tons after it.

Small finches in the Galapagos laid more eggs than ever and raised more young because the rains brought a burst of plants and insects to the usually semidesert islands.

Now the fragile web of living things torn apart by El Niño is weaving together again, adjusting to changes, returning to normal or finding new "normals." But while this destructive "child" was

◀

At the port of Gallao in Peru, fishing boats wait for a run of anchoveta, which once was the largest catch in the world. (FAO PHOTO BY S. LARRAIN)

here, climatologists learned a lot about how the earth reels and readjusts to changes in winds and currents. They discovered some of the clues to predicting the severity of future events like El Niño. Even though we can't do anything about the weather, it helps to be able to understand it. Dr. Richard Barber points out that a society must learn to roll with the climate.

"It's important to absolutely accept that we'll have such an El Niño on an average of once every 7.2 years. It's not cyclical, but only an average frequency. But it's important for an industry, for example, who takes out a ten-year development loan, or a fisherman, or a farmer to know that in one out of seven years something will go wrong, and in one out of twenty years something will go wrong in a big way.

"When people have invested heavily in seed, fertilizer, and pesticides, why not have early warning systems? Knowing ahead, farmers might cover crops. We might draw down reservoirs to handle excessive rains and flooding. We've got to learn to go with nature.

"We're in the Dark Ages when it comes to climate variations. El Niño must have looked like the end of the world in some places. But if we have reasonable explanations, if we have data that tells us we'll be out of this in eighteen months, we have more confidence. Hurricanes, for example, are no longer as demoralizing as they once were because people get explanations. They have more confidence."

Predicting an El Niño with its topsy-turvy effects has obvious benefits. If the blustery blizzard of 1976–77 in the eastern United States could have been predicted, it might have been possible to avoid some of the deaths and suffering brought on by low fuel and food supplies. With early prediction, Africa and Australia might have made preparations against the horrible effects of drought.

The hundreds and hundreds of water samples collected on the 1985 *Wecoma* cruise around the Galapagos provided data on the of nutrients at various levels of the sea. When all the water temperatures were analyzed in the computer and graphed, they showed the depths of the various currents and whether they were welling up or changing direction. Such detailed, tedious collection of data is the basis of research. In this case it will allow scientists to watch and predict the development of the next El Niño.

Dr. Rasmusson says, "The hope is that if we understand the ENSO, we will understand a lot more about how the atmosphere-ocean-climate system works."

One of the things that clouded and misled predictions of the 1982–83 El Niño was the eruption of the Mexican volcano El Chichón in the spring of 1982. Enormous amounts of ash and dust flew into the atmosphere, confusing the satellite sensors and making the temperature readings in the Pacific unreliable.

El Niño taught us a lot about the connection between ocean currents and climate, and El Chichón turned out to give us the same kind of on-the-spot education about the teleconnection of volcanoes.

6

THE VOLCANO CONNECTION

The stillness on Sunday morning, May 18, 1980, was spooky, as though the entire state of Washington were holding its breath. Not even the chirp of a bird echoed over Mount St. Helens. Then at 8:32 A.M., the mountain exploded. The blast was 500 times greater than the 20-kiloton atomic bomb dropped on Hiroshima at the close of World War II. Monster fireworks belched from the volcano. Hurricane-force winds threw millions of 200-year-old fir trees 1,500 feet into the air as though they were matchsticks. An avalanche of rocks and lava roared down the mountain, melting the glacial ice and creating a seething cauldron of boiling mud that raced over farms and towns. More than a cubic mile of volcanic material was thrown into the air, more than a ton for every person on earth.

Dust as fine as talcum powder sifted over everything. Thousands of miles of roads were closed. Air traffic stopped. The town of Yakima in central Washington was buried under 800,000 tons of ash that was blown eastward by the 85-mile-an-hour winds.

When a NASA reconnaissance plane collected air samples over the volcano, they found clouds 63,000 feet into the stratosphere, where they had never encountered clouds before. The choking ash hovered in the air like a gray veil, darkening the sky as though the sun had gone into an eclipse.

By geologic standards, the eruption of Mount St. Helens was "puny." Although a layer of ash from the volcano circled the earth for at least two years causing dramatic red sunsets, the event was a "real dud" to scientists who hoped to find out exactly how volcanic ash changes the climate. The "dud" label had nothing to do with the power of the eruption. If that power could have been harnessed, it would have provided electrical energy for the entire United States for five weeks. The scientists wanted to study the sulfur in the volcanic ash because that sulfur becomes sulfuric acid. In the stratosphere there is no rain to wash the sulfuric acid out. Those acid particles block the sunlight, which results in a cooling of the lower atmosphere.

◀ *When Mount St. Helens erupted on May 18, 1980, the blast was 500 times greater than the 20-kiloton atomic bomb that destroyed Hiroshima.*

(U.S. GEOLOGICAL SURVEY)

The magnificent forests north of Mount St. Helens looked like piles of matchsticks after the eruption. Millions of trees and miles of woodland were destroyed.

(NEIL MOMB)

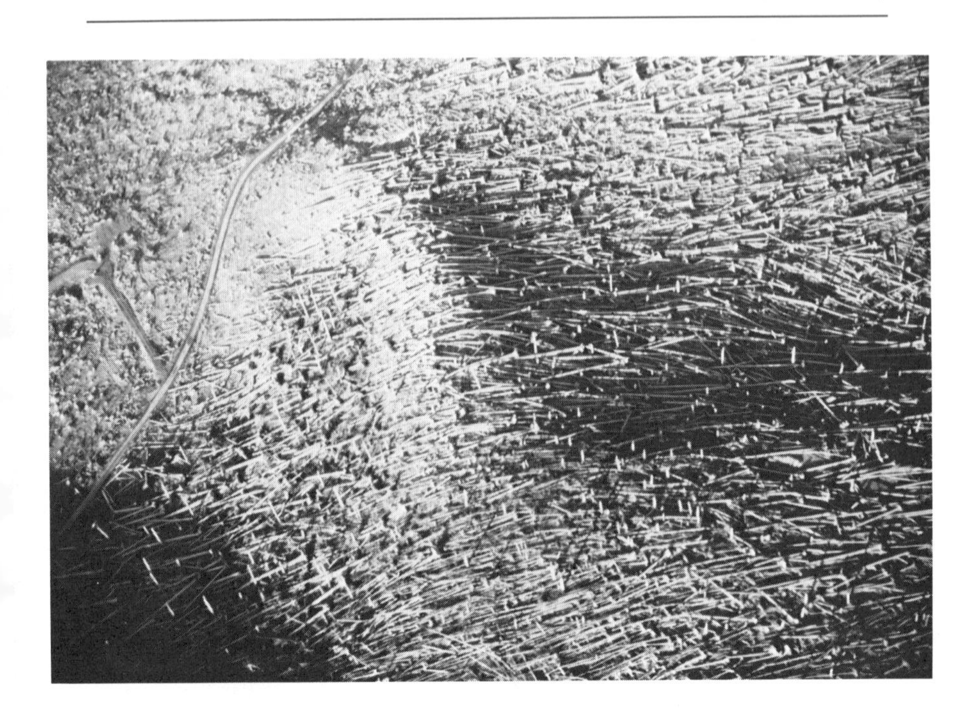

Volcanic ash buried everything after the eruption of Mount St. Helens. Mushrooms like these were the first signs of life to poke through the thick gray layer of dust. (NEIL MOMB)

Because Mount St. Helens exploded sideways, relatively small amounts of ash were blown high into the stratosphere, and little acid remained to have any long-lasting effect on the climate. The aftermath of Mount St. Helens was not like the strange summer that followed the eruption of a volcano called Tambora in Indonesia in April 1815.

People wore winter coats to the Fourth of July parades in 1816. It became known as "the year without summer." In New England a foot of snow covered the ground in June, and in Virginia it was so cold that crops failed. Thomas Jefferson applied for an emergency bank loan to pay his farm bills when his crops blackened from frost. Newspapers carried long descriptions of the events of that strange weather. A June issue of the *North Star* of Danville, Vermont, reported, "The wind during the whole day was as piercing and cold as it usually is the first of November and April. Snow and hail began to fall about ten o'clock A.M., and the storm continued until evening . . ."

On October 31, 1816, a newspaper called *The Messenger* of Zanesville, Ohio, reported that "the number of Emigrants from the eastward the present season far exceeds what has ever before been witnessed." Many New England farmers who had been thinking of moving west decided to leave rather than face another season of failed farms. Records from Maine and Vermont show twice the number of people moved west in 1816 than in any other year in that decade.

The cause of this lost summer in North America and Europe was the eruption of a volcano on the island of Sumbawa in the Dutch East Indies, which is now called Indonesia. Mount Tambora threw some 48 cubic miles of dust, ash, and lava into the air and killed more than 10,000 people in an hour. Crops were buried by ash, and 82,000 more people died of starvation in the famine that followed.

Debris fell on the ocean and coated the water with a thick slush of pumice. Four years later, ships at sea encountered islands of this floating pumice. Lighter ash and dust carried by winds circled the earth in the high stratosphere for months and hung over the Northern Hemisphere in a blanket of insulation that reflected sunlight back into space, reducing the amount that reached the earth.

There is no record showing that anyone at the time made the connection between the eruption of Tambora and the following cold summer, even though thirty years earlier Benjamin Franklin had described the same kind of summer following a volcanic eruption. Franklin may have been the first person to see the connection between volcanoes and climate. When Mount Laki erupted in Iceland in June 1783, Franklin was an ambassador to the Court of France after the Revolutionary War. "During several of the summer months of the year 1783," Franklin wrote, "when the effect of the sun's rays to heat the earth in these northern regions should have been greatest, there existed a constant fog over all of Europe and a great part of North America." He had the idea that the unusually cold winter and strange summer might have been caused by the "vast quantity of smoke" from the Iceland volcano that might have been "spread by various winds over the northern part of the world."

Climate historian H. H. Lamb found an eighteenth century journal from England that describes the summer of the Laki eruption as "an amazing and portentous one, and full of horrible phenomena." The author, Gilbert White, said it was so hot "that butcher's meat could hardly be eaten on the day after it was killed; and the flies swarmed so in the lanes and hedges that they rendered the horses half frantic, and riding irksome."

The eruption of Laki went on for eight months, and various geologic surveys estimate that lava flowed over 220 square miles of Iceland. Volcanic ash must have been thrown high into the stratosphere in order for it to have been spread across Europe to Siberia and into Africa and North America. Sulfurous gases killed crops and pastures in Iceland. Along with half of the livestock, some 10,000 people died from the eruption itself or during the famine that followed. As far away as Europe, people complained of stinging eyes and the rotten-egg smell of sulfur.

When El Chichón exploded in Mexico in the spring of 1982, climatologists, geologists, meteorologists, and all the other scientists involved were ready and waiting. The volcano's name means "lump on the head," and as one meteorologist put it, scientists were willing to accept their own "chichóns" if their research proved faulty on this one. They had been ready for Mount St. Helens, but she didn't turn out to have the right combination of elements to teach them what they needed to know about the climate connection.

El Chichón did. It wasn't the most enormous eruption, in fact not much greater than Mount St. Helens, but the stratospheric cloud it created was a hundred times denser and richer in sulfur. Immediately it was tracked and sampled by every kind of equipment available. The Mauna Loa Observatory in Hawaii measured it with Lidar (an acronym for Light Detection And Ranging), which works something like radar. It sends a laser-light pulse and uses a telescope to measure the timing and intensity of light scattered back from particles in the atmosphere. Ground measurements showed that direct solar radiation was reduced by more than 20 percent, the largest ever seen there. El Chichón's cloud was scanned and photographed by weather satellites, research airplanes, and ground-based cameras. Particles were collected by high-altitude balloons within the main volcanic cloud, and these were later examined with an electron microscope.

The haze from El Chichón is expected to scatter sunlight back into space for several years until the sulfuric acid droplets settle out. The extremely cold winters of 1984 and 1985, when Indiana, Ohio, and other Midwestern and Northeastern states were buried in snow were caused, in part at least, by this haze from El Chichón.

One of the surprises in the study of El Chichón so far has been the discovery that the aerosol cloud may have had its greatest effect on temperatures in the Northern Hemisphere two months after the eruption. That's far sooner than most computer climate models had predicted. A second decrease in temperature followed about a year later, and then it seemed to take longer for temperatures to slowly recover. When a volcano erupts in the Southern Hemisphere, there is a delay of seven or eight months before the Northern Hemisphere's temperatures change. Right after the eruption, the temperature in the stratosphere directly above the equator rose by 38 degrees Fahrenheit (4 degrees Celsius), the warmest ever recorded since 1958, when scientists began taking continuous measurements in the stratosphere.

Most large eruptions have been followed by a few years of cooler climate or erratic weather. Geologists are looking for evidence to show that the aftereffects of volcanic action last much longer, for decades or even centuries.

Debris from volcanic ash and dust, along with other samples of the earth's ancient atmosphere, are locked in deep layers of ice. International teams of geologists and climatologists are discovering the secrets of past ice ages in snow that fell when cavemen were drawing images of prehistoric animals on cave walls. Not only can they reconstruct factors of ancient climates, but they are finding clues that may help predict future changes as well.

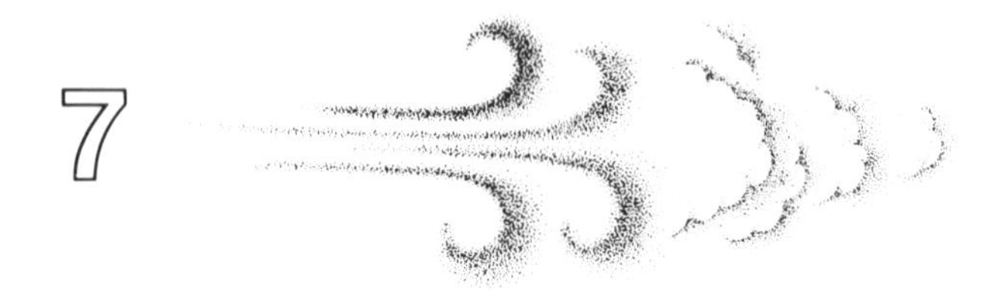

7

SECRETS IN THE ICE

Scientists bundled in layers of down jackets, caps, and mittens may look as though they're out to build snowmen, but they are on their way to work in ice-walled laboratories carved in a glacier. One international team of geologists at Camp Century, Greenland, is drilling deep into the ice to find out what the earth's atmosphere was like 100,000 years ago.

"Ice is nice," says Dr. Chester C. Langway, Jr., chairman of the geology department at the State University of New York at Buffalo. "It's a permanent record of particles of volcanic dust, trace metals, sea salts, anything that was in the atmosphere."

Such particles also drift and settle in the sediments at the bottom of the oceans, but there it's not quite as accurate a record. "It takes a thousand years for a few inches of ocean sediment to accumulate," says Dr. Langway, "but it is so easily stirred up that it doesn't stay in neat chronological order as ice does."

Like firewood stacked ready for burning, hundreds of long metal canisters of ice ready for study are stacked in cold-storage vaults in the Ice Core Laboratory, where they are kept at minus 30 degrees Fahrenheit (minus 35 degrees Celsius). Dr. Langway established this "library" of ice in Buffalo in 1975 as the nation's central bank for all the ice-core samples drilled in Greenland, Antarctica, or anywhere else in the world. The ice is cut into three- to five-foot lengths for convenience because even at that size each weighs about forty pounds.

Sherri Rumer (above) *prepares ice-core samples for chemical analysis in the laboratory at the State University of New York at Buffalo. Dr. Hitoshi Shoji and Dr. Susan Herron* (below), *working in down jackets and warm gloves, prepare an ice core for cutting in the same laboratory where all ice-core samples are stored.* (ED NOWAK, NEWS BUREAU, SUNY BUFFALO)

The cores are available for any scientist to study, and in one year there were more than four hundred requests for longitudinal slices of ice from specific years and locations.

At the Camp Century base, Dye-3, Greenland, geologists from America, Denmark, Japan, and Sweden eat, sleep, work, and relax in a community built on stilts. The building, which has to be jacked up higher every few years to keep it free of snow, is also a radar base that once was part of the Distant Early Warning system built in 1960. It looks like the kind of drilling platform used by crews looking for offshore oil and gas. Even in the summer, when most field researchers work at the station, temperatures in the drilling trenches average between 14 to minus 4 degrees below zero Fahr-

Dye-3, Greenland, the U.S. communication site on the ice sheet in southeast Greenland. The top of the radar dome is as high as an eight-story building. Drilling to bedrock, some 6,700 feet below the surface, was accomplished in 1981. A continuous four-inch-diameter ice core was recovered for scientific study.
(C. C. LANGWAY, JR., ICE CORE LABORATORY, SUNY BUFFALO)

enheit (minus 10 to minus 20 degrees Celsius). A single day's drilling operation may last twenty-four hours.

The first continuous ice core at Camp Century was drilled through a polar ice sheet four-fifths of a mile to bedrock with a special drill built by the Denmark team. The core, which is four inches in diameter, is cut into the three- to five-foot lengths that can be easily handled. At the Byrd Station in Antarctica, another team recovered an ice core from a mile and a third to bedrock.

Ice begins to change from decompression as it nears the surface, and the teams have to work fast taking the first tests and measurements. Working in the "clean rooms" of the ice-walled laboratories inside the glacier, they pack ice cores in metal canisters and ship them to the Buffalo storage lab.

Ice-core drilling had its start in Alaska in 1956, when the Army Corps of Engineers was studying permafrost, which is that layer of soil that never defrosts; it's always frozen. Building on the techniques learned in those studies, the present ice studies have grown into an international project funded by the Division of Polar Programs, part of America's National Science Foundation.

"Each year, on the high polar ice sheets, new snowfall buries the previous accumulation deeper, increasing the density of the underlying granular snow," says Dr. Langway.

Trapped in these layers of snow are bits and pieces of fallout from the atmosphere, which may have come here from the sea, from space, from volcanoes, from plants or animals, or from man-made things. Air is trapped, too, and when that air is drawn off in the laboratory, it reveals how much carbon dioxide was in the atmosphere at various times. Today's level of carbon dioxide is 331 parts per million. In ice formed before the Industrial Revolution, they found the carbon dioxide was 275 parts per million, but ice laid down during the Ice Age shows only 200 parts per million. The rise in the carbon-dioxide level after the Industrial Revolution is understandable, but why was the level of carbon dioxide so low during the Ice Age? Nobody knows yet, but scientists trust that when they do find out, the answers will help us deal with future problems.

Snow, rain, or any kind of precipitation is affected by what's in the atmosphere when it falls. Dr. Michael M. Herron, who also works at the Ice Core Laboratory in Buffalo, says, "Once it lands on the ice sheet or ice shelf it remains frozen, so you have a general record of the composition of the atmosphere over time spans as long as 120,000 years, and up to perhaps several million years in portions of East Antarctica."

One of the important things this research shows is what's really "normal" for our atmosphere. A few years ago there was a big scare when fish were found to contain dangerous levels of mercury, apparently from the water. But nobody knew what a normal mercury level was for water. How much mercury did fish pick up from unpolluted water? How high were the mercury levels a hundred years ago, or a thousand? With ice-core samples, it's possible to find out about the chemical and physical composition of the atmosphere and water before man's interference. And that can't be done by any other means right now.

Ice-core specialists can distinguish between particles of coal and wood. They can study the carbon particles in glaciers to trace the history of burning. They can study lead aerosols, dusts, and hydrocarbons to find minute quantities of elements such as aluminum, cadmium, chlorine, and potassium that will provide a natural level or base line against which they can measure the effects of mankind's pollutions.

Plutonium, which is a fallout from nuclear testing, is also evident in ice. Core samples showed a peak in the plutonium level in ice laid down from 1946 through 1948 when we detonated atomic weapons at the testing grounds in Almagordo, New Mexico, and when we bombed Hiroshima and Nagasaki. Fallout from aboveground nuclear weapons testing in the late 1950s and early 60s shows up in ice layers, and so does plutonium from 1964 when a navagational satellite was launched. The rocket didn't boost the payload into orbit, but came down in the Indian Ocean, burning its generator during reentry.

Because they can easily date recent events, geologists know that their tests for reading the deeper, older layers of ice are accurate, too. They are so accurate that annual layers laid down in the last

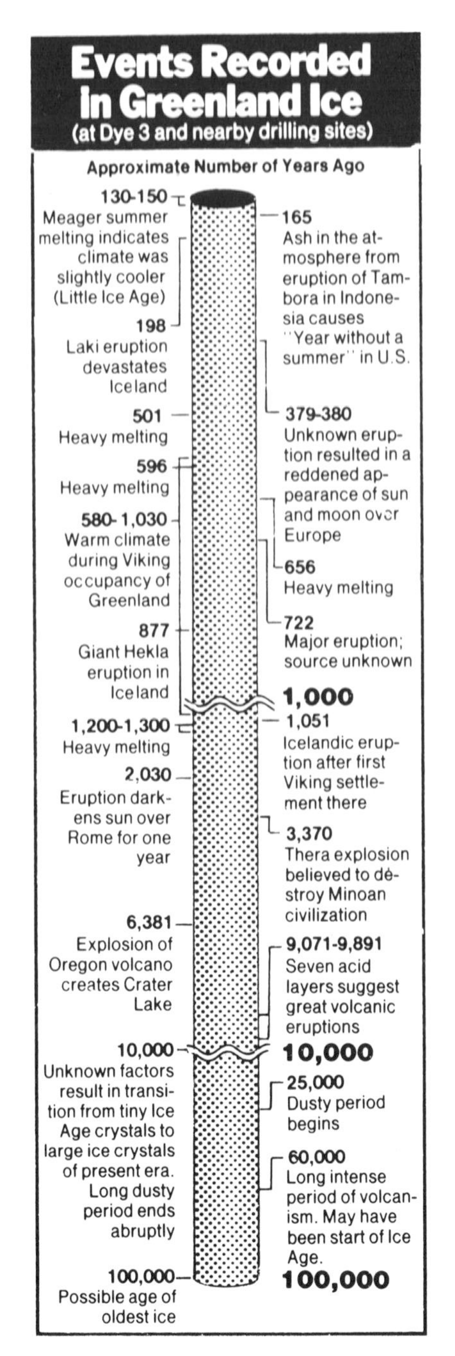
Events Recorded in Greenland Ice
(at Dye 3 and nearby drilling sites)
Approximate Number of Years Ago
130-150
Meager summer melting indicates climate was slightly cooler (Little Ice Age)
165
Ash in the atmosphere from eruption of Tambora in Indonesia causes "Year without a summer" in U.S.
198
Laki eruption devastates Iceland
501
Heavy melting
379-380
Unknown eruption resulted in a reddened appearance of sun and moon over Europe
596
Heavy melting
580-1,030
Warm climate during Viking occupancy of Greenland
656
Heavy melting
722
Major eruption; source unknown
877
Giant Hekla eruption in Iceland
1,000
1,200-1,300
Heavy melting
1,051
Icelandic eruption after first Viking settlement there
2,030
Eruption darkens sun over Rome for one year
3,370
Thera explosion believed to destroy Minoan civilization
6,381
Explosion of Oregon volcano creates Crater Lake
9,071-9,891
Seven acid layers suggest great volcanic eruptions
10,000
Unknown factors result in transition from tiny Ice Age crystals to large ice crystals of present era. Long dusty period ends abruptly
10,000
25,000
Dusty period begins
60,000
Long intense period of volcanism. May have been start of Ice Age.
100,000
Possible age of oldest ice
100,000

38

nine centuries identify dates of volcanic eruptions within a year, and in ice fourteen centuries old they can label the action of volcanoes as close as three years.

"From the study of the cores," says Dr. Langway, "scientists have reconstructed global volcanic activity over the last 10,000 years."

Records from Europe in 1601 and 1602 describe a faint reddish sun and moon. Ice samples from rain or snow that fell in 1601 and 1602 show peaks of acid deposited from large volcanic eruptions. When Julius Caesar was assassinated in 44 B.C., there was a dimming of the sun. Ice cores tell why. In Greenland, about 50 B.C., a volcano erupted with a heavy fallout of sulfuric acid that stayed in the stratosphere like a curtain over the sun and moon for many years.

Another thing that helps geologists count the annual layers of ice and describe the climate is the summer melting. The layers that show summer meltings, for example, match the warm years between 950 and 1400 when the Vikings had settlements in Greenland.

The great ice masses of Greenland in the Northern Hemisphere are the leftovers of enormous glaciers that slowly ground their way across North America more than 60,000 years ago. Even though these appear to be huge sheets of ice, they are small compared to the ice sheets that covered about 30 percent of the planet as recently as 18,000 years ago. By measuring the advance and retreat of these smaller ice sheets, glaciologists discover how climate both controls and is controlled by the ice masses. Similar studies have been especially effective on the Antarctic ice sheets.

It's impossible to measure every glacier on earth, so detailed measurements are made on a few. From that information, it's possible to predict how any ice sheet works under known conditions. Some measurements are taken historically. Photographs or drawings of glaciers in the Alps, for example, show changes at the front. A ridge called a moraine, made up of boulders and other debris pushed ahead of a glacier, is left standing when the river of ice retreats by melting. These moraines are easily dated. This retreating motion of glaciers is also called backwasting.

To measure the loss or gain of ice build-up, geologists hammer wood or metal stakes into the surface of glaciers and at the front of them. In some ice fields transmitters have been planted to send signals that will be picked up by satellites, making it easier to get results because then geologists don't have to hike to the glacier to measure the stakes.

In the 1950s several aircraft crashed on the Greenland ice sheet because their radar didn't work. The ice was transparent to the frequency used by the radar and didn't bounce back any signals. These tragic accidents, however, resulted in the use of this radar as a device for studying ice. Now it is used to take detailed pictures of the layering of ice and of the contours of the earth buried beneath it.

The activity of a glacier is a measure of climate. In southern Alaska, for example, the Columbia Glacier and others like it may move twenty feet a year. Although a lot of snow falls and accumulates on these glaciers, they move rapidly in the relatively warm climate. In the colder regions of the high Arctic, glacial ice melts slowly, so that the glaciers move slowly, perhaps only three feet a year.

As a glacier retreats, it loses ice on the surface while it also moves forward like a conveyor belt dumping debris. Since ice is a good insulator, deep inside an ice sheet the temperature is higher than at the surface. Therefore, ice melts at the bottom as well, forming a lubricating layer of water that lets the ice move. When the inner core of the ice sheet thins and melts, the outer edge advances. Where glaciers reach the sea, this motion results in calving as great chunks of ice break off and float away as icebergs. In Antarctica, which has more than 90 percent of the world's ice, the ice sheet becomes so thick and the temperature at the base gets so high that great chunks of ice continually slide into the sea. No one is certain just how much ice is lost by calving or from surface melting of the floating Antarctic base that extends over an area of 800,000 square miles.

Glaciers perpetuate themselves, and even small glaciers have an immediate effect on weather. On the Greenland ice sheet, it is much colder than at the same latitude in Siberia or Arctic Canada because the enormous white surface loses radiation energy as it is reflected back into space. When the huge ice sheet of Antarctica builds, the albedo effect increases, and that results in a general cooling of the earth's climate. The colder climate, of course, adds to the building power of the ice sheets.

The Columbia Glacier in Alaska shrunk 3,608 feet in 1984, almost twice the amount of the previous year. The U.S. Geological Survey has been monitoring the glacier and predicts that it will change the shape of Alaska in the next ten years as it uncovers a long, deep fjord. Icebergs are breaking off this river of ice four times faster than in the 1977 calving. (U.S. GEOLOGICAL SURVEY)

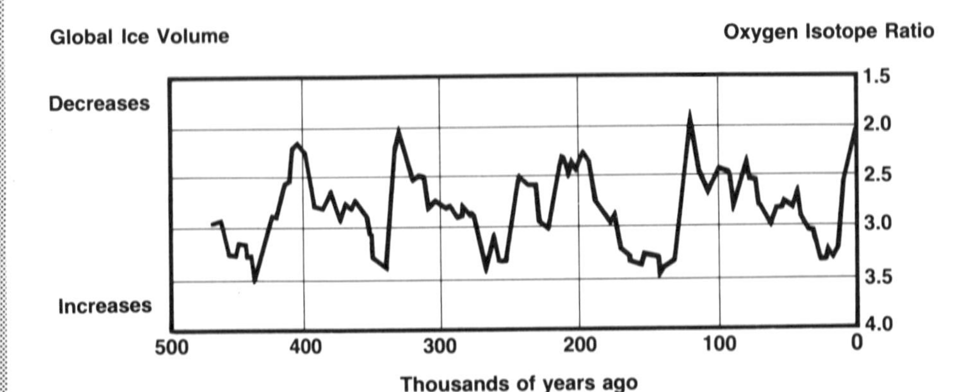

Ice ages tend to occur every 100,000 years. The peaks on this chart represent periods of less ice on the earth (interglacials). We are at or near the end of one of these warm periods, and we will begin moving toward increasingly colder climates if the pattern continues.

With the concern about an increased greenhouse effect warming the earth, glaciologists have increased their efforts to find out how and why ice melts, as well as how and why it changes climate.

The science of glaciology, according to one scientist, is just getting interesting and just beginning to ask the right questions. The earth's climate during the past two million years featured ice, and we live now in an unusual time relatively free of ice. According to records from ice cores, there have been about eight interglacial periods each lasting 10,000 years. We're in one now, one that is coming to an end, but also one that is different from all the others because we *do* live in it.

8

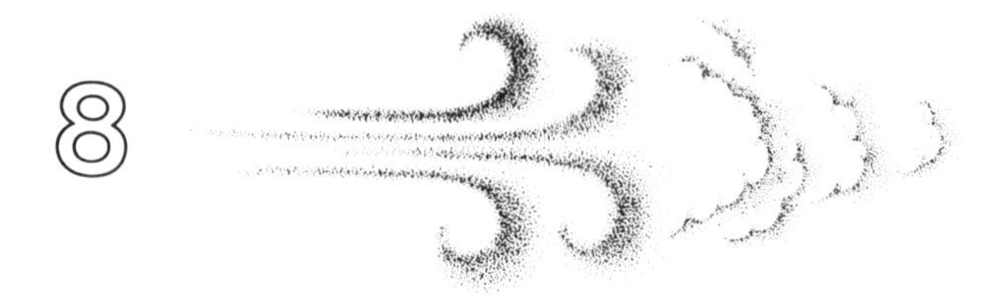

THE FOREST CONNECTION

A stately looking llama named Ferrous, with chartreuse ribbons flying from his halter, marches through the Adirondack Mountains of New York State behind a team of scientists looking for evidence of acid rain. The bright ribbons are signals to deer hunters that Ferrous is not fair game.

Llamas have worked for the United States Forest Service in Colorado, and they seem to be perfect pack animals for the Adirondacks as well. They eat almost anything, their thick wool keeps them comfortable in all weather, and they easily carry heavy loads over narrow, steep trails where trucks, cars, or even horses would find the going difficult.

Four research teams will spend three years sampling the soil, water, and fish in the 2,759 ponds and lakes of Adirondack Park. They will find out exactly how much damage has been done by the acid rain that is destroying forests as surely as bulldozers or the axes of settlers.

In the early days of America, forests were so thick that it was said a squirrel could leap from tree to tree and never touch ground from one end of the thirteen colonies to the other. If you travel that route now, following the super highways that connect those states along the East Coast, you might wonder where the forests went.

The forests were something to write home about, and the early settlers did. They also wrote about the beautiful weather in their New World home. But even as they extolled the wonders of the trees and climate, they set about clearing the land for towns and farms. It certainly must have looked like an endless supply of trees. Probably not a voice was raised against such progress. Open land was what they wanted to make travel easier and to provide some safety from then-hostile Indians and dangerous predators like bears and mountain lions.

In a history of climate change in America, geographer Kenneth Thompson from the University of California said, "The result was, of course, the wholesale destruction of great virgin forests at a scale and speed never equaled elsewhere in the annals of environmental history."

Running through letters back home to England and France were accounts of a warming trend in the weather. At a meeting of the American Philosophical Society in 1770, Dr. Hugh Williamson read a paper called, "An Attempt to account for the Change in Climate which has been observed in the Middle Colonies of North America." He reported that long-time residents had noted during fifty years a "very observable Change of Climate" with milder winters and cooler summers as a direct result of forest clearing.

Most people believed in a forest-climate connection. Columbus was certain that the afternoon rains in the West Indies were produced by the lush forests, which was why they were known as rain forests. His son, Ferdinand, wrote an account of his father's voyages, saying that Columbus knew "from experience" that mist and rain had been reduced in the vicinity of the Canary Islands, Madeira, and the Azores because great tracts of forests had been removed.

Thomas Jefferson believed there was a connection between forests and climate, too, although no one really knew what that specific connection was. One of the first scientists to disagree was the German Baron Alexander von Humboldt (after whom the Humboldt Current along the coast of Peru was named). After he visited America, Humboldt pointed out that without accurate measurements there was no proof that forests had any effect on climate.

In the 1850s another scientist, Lorin Blodget, suggested that climate was permanent and not affected by anything under man's control, such as the removal of trees or cultivation of land.

So the debate continued, based only on observations but no hard data. People believed what they saw, and they saw changes in weather as forests were destroyed. In 1867 the New York State Agricultural Society issued a report saying that the adverse climate was the direct result of cutting down forests. Some people began thinking that it might be a good idea to plant trees on the treeless Great Plains because if cutting down trees could cause lack of rain, then planting trees could bring more rain.

Then in 1871, the worst forest fire in American history wiped out more than a million acres of timber and killed 1,500 people in the town of Peshtigo, Wisconsin. The disaster focused attention on the forests and probably helped push through 1873 federal legislation to promote the planting of trees for rainmaking. A few years later, the American Forestry Association was formed, and that led to the Division of Forestry as part of the United States Department of Agriculture, which is the basis of our present National Forest system.

The climate-forest connection is probably one of the first big issues that made people aware of their responsibility to the environment. You don't have to be in a forest long before you feel the buffering effect of the trees against the wind, or the change in temperature, but it took until the 1980s to find evidence that showed *how* the forest-climate connection works.

Although some of the deforestation began with the settlers sent out to colonize the world, it moved into high gear with the Industrial Revolution. England had so many merchant ships on the high seas in the 1800s that the great wooden sailing vessels loaded with raw cotton from India would hail the ships carrying finished cotton cloth back to India as they rounded the stormy tip of Africa. Britian's cotton exports doubled every twelve years, and someone estimated that by 1850 they were shipping enough cloth to make a shirt for every person then on the planet.

Shipyards boomed, not only with the merchant vessels, but also with the great fleets of the Royal Navy. As a result, the magnificent

A 1935 print of Brazilian forests being cleared for a coffee plantation. Despite steady deforestation over the years, the threat of serious global tropical rain forest depletion has come within the last ten years.

(EARTHSCAN/MARCOS SANTILLI)

golden oak forests of the British Isles were almost depleted by the end of the 1800s. England began importing fine-grained, insect-resistant timber from the lower Himalayas and hard teakwood from Burma and West Africa.

It was also the era of enormous plantations. Thousands of acres of lush monsoon forests in Burma, Thailand, and the Philippines were cleared for growing rice, coffee, indigo, tobacco, sugar, and other export crops. By the end of World War II, the sugar cane plantations of the Caribbean had replaced all sizable forests. And it hasn't stopped.

Two-thirds of Latin America's original forests are gone or are seriously depleted. The United States Office of Technology Assessment estimates that 15,000 acres of timber in the Amazon basin are being wiped out *each day* to supply lumber companies in the United States and Europe. In the Philippines and New Guinea, thousands of acres of forests are cut each year for cheap paper pulp for world-wide printing industries.

The rain forests of Central America are not being destroyed to feed the hungry people of the Third World nations but to supply the fast-food restaurants of the rich countries. In what has been called "the hamburger connection," nearly 40 percent of the forest cover of Central America has become pasture for cattle. Ninety percent of the cattle are exported to America, where they are sold to the fast-food restaurants.

The coca plant thrives on the steep, eroded hills left when forests are cut, especially in Peru, Columbia, and other South American countries. Cocaine is made from coca, and it's such an easy and profitable crop to raise that landowners aren't likely to opt for the long, slow process of replanting forests.

Half of Africa's woodlands have been leveled, and most has been haphazard clearing for crops or grazing. The jungles have been replaced with virtually useless shrubs or sharp-edged grasses useless to grazing animals.

Loading teak logs in Thailand (FAO PHOTO)

An area of western Brazil burned for farming. Here an influx of settlers has turned once lush forests into a wasteland of erosion and brick-hard soil.
(EARTHSCAN/MARCOS SANTILLI)

Because half of the world's people use firewood for fuel, areas around settlements and towns are stripped. About 90 percent of the wood in Africa, two million hectares a year, is cut for fuel by people struggling to live on the edge of starvation. (One hectare is equivalent to 2,471 acres of land.)

In more industrialized nations, the stripping is being done, as in the Adirondacks, by acid rain. It's not a new phrase and not a new problem. The term "acid rain" was coined by an English chemist, Robert Angus Smith, in 1872, when he noticed the rain in the industrial city of Manchester was more and more acid.

Acid rain is not a temporary problem, nor is it a minor one. It is a major global concern that threatens to destroy forests everywhere. Started in factory towns, the pollution of acid rain is carried everywhere with the winds. It is found in Brazil, China, South Africa, and even the world's last wilderness, the Arctic and Antarctic. Half of Germany's famous Black Forest has been damaged. In Switzerland the destruction of trees could prove dangerous because belts of trees above Alpine villages are protection against avalanches and landslides.

Almost nothing is safe from acid rain, not plastics, metals, glass, or stone. In 1979 the six marble statues of the maidens called the Caryatids on the ancient Acropolis in Athens were replaced by cement replicas. They had stood for 2,500 years, but the effects of pollution were turning the original stone to soft gypsum. The Colosseum in Rome, the Taj Mahal in India, and Cologne Cathedral in Germany are among the treasured buildings constantly undergoing repair from acid-rain damage.

When fossil fuels burn, the waste products released are sulfur dioxide and the nitrogen oxides. Combined with water and oxygen, these gases become sulfuric and nitric acids. By comparing samples of rain frozen deep inside the Greenland icecap before industrial times, scientists know that this acid rain and snow is forty times more acidic than "normal" rain. There are natural causes of acid rain: volcanoes, forest fires, even lightning. But there are towering smokestacks on factories and smelters that puff out twice as much sulfur every year as Mount St. Helens spewed out at its most active time.

Acid rain works in many ways to destroy forests. It leaches out vital nutrients from the soil, replacing them with a toxic "cocktail" of heavy metals. It stunts root growth, etches leaf surfaces, causes genetic changes, and leaves trees unable to reproduce.

In spite of accumulating evidence, not everyone agrees that acid rain is a major problem. In December 1984, the *New York Times* reported, "The Environmental Protection Agency has rejected a petition by New York and other states to curb acid rain. The Administration is not convinced that gases from the coal-fired power plants of the Ohio Valley are the cause of the acid rain that falls hundreds of miles away in the Northeast and Canada."

Canada has requested action against the acid rain sent northward from America's industries. America now finds itself making the same request of Mexico. The winds blow north, carrying pollutants with them.

For obvious reasons, factories cannot be shut down just to see what happens to acid-rain levels. But a different kind of experiment in the Rocky Mountain states has provided a clear link between the cause and effect of acid rain. There, the main source of sulfur dioxide gas comes from a handful of copper smelting plants. The emissions from these factories are easy to track because they aren't steadily puffing away, but vary from year to year as the copper market varies. In a study by the Environmental Defense Fund, the records of rain-sampling stations from Arizona to Idaho show distinct changes in acid levels that correlate exactly with the tonnages of sulfur dioxide emitted by the copper smelters, which are upwind of the sampling stations.

"The importance of the new study," says the *New York Times* report, "is evidence of a direct relation between what goes up and what comes down."

There are, of course, many consequences of acid-rain damage to lakes, wildlife, human life, and buildings. But there is also the fact that the destroyed forests will change our climate.

Columbus was right. There is a forest-climate connection. Forests do make rain. There was no hard, scientific proof of this fact until an international group of scientists, including oceanographers, hydrologists, climatologists, and ecologists, used the Amazon River basin as a laboratory. Dr. Eneas Salati, a professor of meteorology at the University of São Paulo, Brazil, headed the team who made its report in 1984. These scientists proved that forests play a vital role in actually generating weather and that a forest can return up to 75 percent of the moisture it receives back into the atmosphere.

One-fifth of all the river water poured into the earth's oceans comes from the Amazon, where two-thirds of all the free fresh water is constantly percolated through a never-ending cycle. ("Free" fresh water is aboveground water as compared with "locked up" water in underground reservoirs and polar icecaps.) Within this

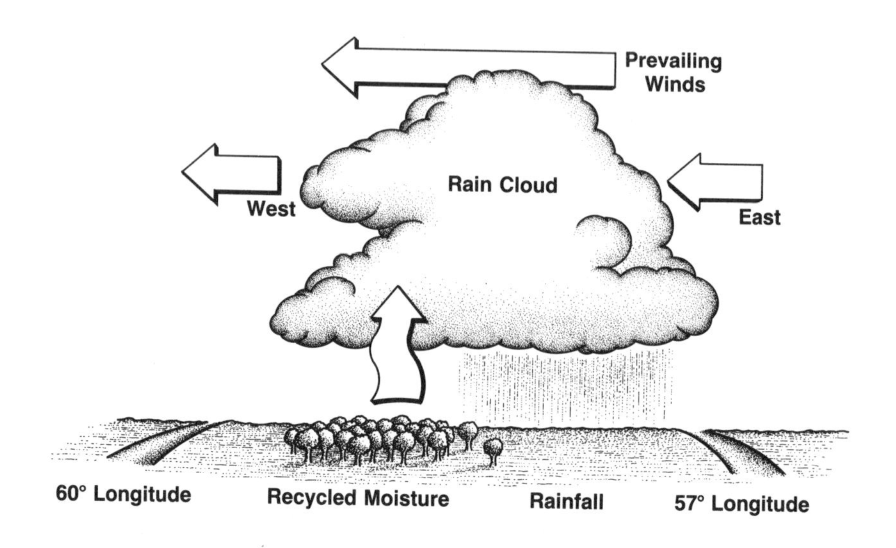

Rain clouds carry moisture picked up over the Atlantic Ocean and move westward over the Amazon forests. Some of the water vapor condenses and falls as rain, but it is partially replaced by moisture recycled back into the air from the forests themselves. When the rain clouds move westward, they are mixtures of water from both the forests and the ocean.

two-and-a-half-million square miles of Amazon rain forest, the scientists collected data to show that much of the water a forest gathers can be returned to the air in amounts large enough to form rain clouds. Land covered by trees collects and returns to the air at least ten times as much moisture as bare land, and twice as much as land covered by grass or vegetation other than trees.

Trees transpire and give off moisture. Water also evaporates from leaf surfaces, so that clouds form above the forests. Water vapor condenses and falls as rain. As the temperature of the land beneath the clouds changes, winds blow, carrying moisture with them and picking up more moisture over the ocean.

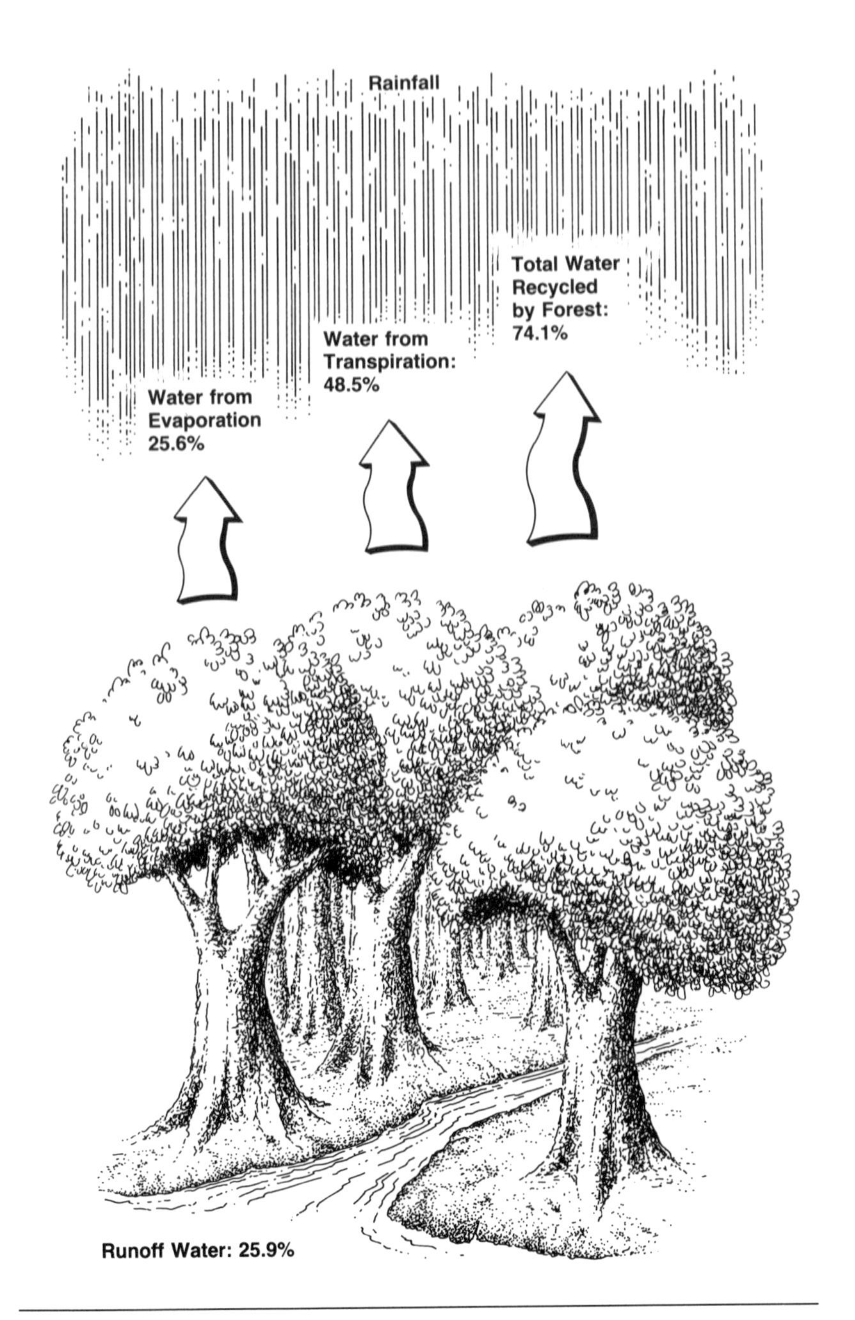

Forests help create the rain clouds that hang over them. Almost half the moisture comes from transpiration, which is water released through pores in the trees, and another 25 percent evaporates from the surfaces of the leaves.

When scientists analyzed samples of air moisture taken in a 2,000-mile east-west line from the Atlantic Ocean, across the Amazon basin, to the Andes Mountains, they found that the forest itself produced 50 percent of the rain that fell on it.

Dr. Salati and his team do not claim to predict what will happen to global climate if these rain forests are destroyed because there are too many variables. But they are certain of the outcome for South America. They know that the destruction will cause major changes in agriculture because there will be less rainfall. They know that with fewer clouds forming there will be a change in the amount of solar heat reflected back into space. Satellites have gathered evidence showing that the earth's reflectivity, or albedo, returns about 30 percent of the sun's energy. A large part of the radiation balance of the earth is determined by the albedo, and that, in turn, is one of the factors regulating the distribution of heat around the planet.

If tropical rain forests in all the equatorial countries around the world are destroyed at the rate of South America's, we are likely to see major changes in world-wide weather.

9

THE SOLAR CONNECTION

The "Ace Satellite Repair Company" went to the rescue and put an end to the "throwaway" satellite. To replace the Solar Maximum Satellite, known as Solar Max, would have cost 235 million dollars, but fixing it in orbit cost only 50 million dollars.

In April 1984, the world watched as the crew of the shuttle Challenger hauled Solar Max into the cargo bay, fixed it, and sent it back into orbit again on the first successful satellite repair mission.

Solar Max was rocketed into orbit in February 1980 to study the solar flares that erupt violently every eleven years on the surface of the sun when sunspots reach a peak. During this stormy season, the sun bombards the solar system with x-rays and with high-energy particles that cause radio blackouts and the colorful auroras known as the northern lights or aurora borealis.

For almost a year, Solar Max monitored the sun. It took pictures of the sun's hot halo, called the corona, and found evidence of thermonuclear fusion in the solar flares. But most important, it sent back messages showing that the sun fluctuated in power. Its energy output dropped by two-tenths of a percent whenever a large cluster of sunspots appeared. That answered the age-old question, "Does the sun change?" But it presented new questions, such as, "How does the sun store the extra energy hidden by the sunspots?"

The sun has been so predictable that the great stone calendar called Stonehenge, built on the Salisbury Plain in England 4,000 years ago, still accurately marks the sunrise on the summer solstice. But even as steady as the sun's schedule seemed to be, scientists wondered about geologic evidence for a pattern of changes showing seven ice ages in the last 700,000 years with each ice age separated by an interglacial period of 10,000 years. These interglacial times were not consistently warm years, but were interrupted by fifty-year spurts of cold climate. What we've come to call the Little Ice Age during the Middle Ages lasted from the 1500s to the 1700s. Some climatologists think we are at the peak of one of these interglacial times now and that the earth is beginning to cool as we move into the next full ice age.

The earth's "normal" climate is a pattern of ice ages sprinkled with short warm spells. Why does the earth follow this rhythm? Are ice ages in some way connected to the rhythms of the sun? A change in the amount of solar radiation reaching the earth might explain these cycles.

Milutin Milankovitch, a Yugoslavian astronomer, thought so. He came up with three ideas in 1930 after he had done an incredible number of calculations of the kind we now feed into a computer. His figures told him that the three relationships between the earth and the sun that dictate the pattern of climate are wobble, tilt, and orbit.

The earth wobbles like a slow-spinning top. The North Pole doesn't always point in the same direction but traces a circle in space that takes about 22,000 years to complete. Winter started in July instead of December in the Northern Hemisphere 10,000 years ago because of the earth's wobble.

The earth tilts. The imaginary poles do not run straight up and down, and this tilt is the cause of seasons as the sun strikes our planet at varying angles. The angle of this tilt changes slowly from 21.8 degrees to 24.4 degrees and back again in a cycle of 41,000 years.

The third relationship that changes our climate is the pattern of the earth's orbit around the sun. From a perfect circle, it flattens out to an elipse and back again every 100,000 years. Right now our orbit is almost a perfect circle. When we are farthest from the sun in the part of the elliptical orbit most flattened out, we get 20 percent less solar energy than we do when we are closest to the sun.

Milankovitch's theory was put on a back burner for many years, but in 1976 some proof surfaced from the deep ocean floor. Geologist J. D. Hayes from Columbia University, oceanographer John Imbrie from Brown University, and climatologist N. J. Shackleton from Cambridge analyzed the sediment cores drilled from the Indian Ocean, much as cores are drilled from ice. They found evidence of climate changes that just about matched Milankovitch's prediction. The dominant cycle was 100,000 years, with two lesser cycles of 42,000 years and 23,000 years.

Long before we connected the cycles of thousands of years to the sun, scientists searched for proof that shorter rhythms might also be caused by solar radiation.

Since the time of Aristotle, people believed the heavens, being made by God, must be perfect and without blemish. In the seventeenth century when Galileo saw spots on the sun, he didn't report them immediately for fear of reprisals from the church. Not only did he see a sun with blemishes, but with blemishes that changed periodically as though some other force were at work.

Sunspots have gone in and out of fashion as a subject of study. The Chinese knew about sunspots long before the western world acknowledged them. A German pharmacist, Heinrich Schwabe, whose hobby was astronomy, was convinced he'd find another planet close to the sun. Although he searched for years, he didn't find a planet, but he did become fascinated by sunspots. In 1843 he published his sketches in a report describing how sunspots increase slowly to a maximum and then decrease to a minimum over a ten-year period. More accurate measurements later showed the time between this maximum and minimum to be eleven years.

This eleven-year cycle has since been used to predict everything from changes in crop yields in Europe, lake levels in Africa, the population of rabbits in Australia, harvests of vintage wine, to the

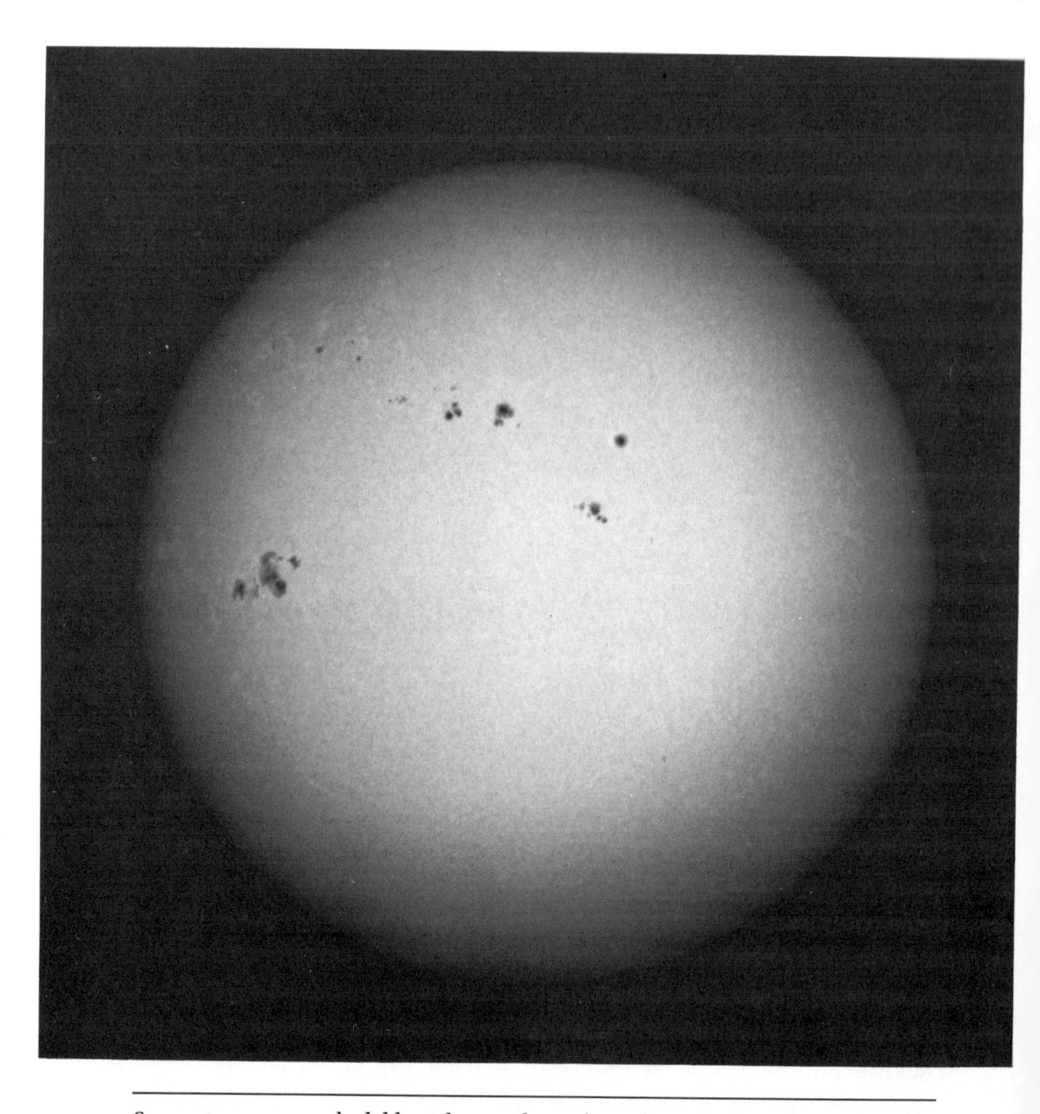

Sunspots appear as dark blemishes on the sun's surface. They are enormous areas cooler than the rest of the sun's surface, and they tend to appear in eleven-year cycles. (DANIEL MARCUS AND WILLIAM SMITH)

stock market in New York. With all this nonscientific interest in sunspots, scientists were skeptical. The whole thing seemed more like myth and old wives' tales, more part of the non-science called astrology. So they did little more than admit sunspots existed until recently.

Sunspots are enormous cooler areas, several times larger than the earth's surface, passing across the sun. They are violent storms of charged particles that create a gigantic magnetic field thousands of times stronger than the earth's, and they block some of the heat from the sun's interior.

At sunspot maximum, the surface of the sun is turbulent with gigantic prominences of gas leaping tens of thousands of miles into space. Like a billion hydrogen bombs, the flares of gas explode, making it look as though the sun were wearing a halo. Solar winds, which are charged particles, travel across space intercepting and being captured by the earth's magnetic field.

Every eleven years at the peak of solar activity, the sun's magnetic field reverses, and it takes just about twenty-two years for the sun to go through a complete magnetic cycle back to its original magnetic position. Climate changes seem to come in cycles of eleven and twenty-two years, most of the time.

When an English astronomer, E. Walter Maunder, searched old records and journals, he found a gap in the eleven-year pattern of sunspots. Between 1645 and 1715 there was little or no sunspot activity, and that time of silent sun, now called the Maunder Minimum, was also the time of Europe's worst climate during the Little Ice Age. Recently at the the National Center of Atmospheric Research, John Eddy found two other periods of minimum sunspot activity, from 1100 to 1250 and from 1460 to 1550. Checking his facts, he compared his information with data collected from tree rings.

Every tree keeps a record of weather history. The first person to connect tree rings to climate was Andrew Ellicott Douglass. He started the Tree Ring Laboratory at the University of Arizona to see if he could match sunspot cycles with tree rings on the assumption that thick rings would mean good growing years and thin rings, poor years. The only problem with that was the difference in trees. Some species are more affected by differences in temperature.

Dendrologists, who study trees, now use more exact measurements by comparing the amount of oxygen-16 to oxygen-18 in a tree ring. Some water molecules contain a "heavy" oxygen called oxygen-18, and others have oxygen-16, or "light" oxygen. It takes more energy to evaporate water containing the heavier oxygen-18 than the lighter oxygen-16. That means that in a warmer year, with more energy from the sun, the water that evaporates from oceans and lakes to fall as snow or rain has a higher proportion of the heavy oxygen than in cooler years. When dendrologists analyze tree rings, they compare the ratio of oxygen-16 to oxygen-18 to get an indication of the average yearly temperature. The same method is used to analyze ice-core samples. In colder weather, the precipitation is less likely to be water containing oxygen-18. For that reason winter snow contains relatively little oxygen-18.

A seventy-two-year sequence of rings from giant sequoia trees shows an eleven-year sunspot pattern, and the ancient bristlecone pines have shown a similar cycle in the last 7,000 years of growth. Midwest America seems plagued by drought every twenty years. Tree rings taken from forty different sites around the Midwest show a twenty-two-year cycle that coincides with the middle of the droughts and the alternate pattern of sunspot activity.

All these studies show that the sunspot cycles seem to be more than coincidence when matched with our changing climate. The Solar Max satellite sent back some measurements of solar radiation that indicated the sun's radiation increased one or two tenths of a percent during peak sunspots. The connection to climate may be simple. The sun gets hotter as more sunspots appear, and as the sun gets hotter, the earth gets slightly warmer, too. When sunspots decrease, both the sun and the earth get cooler. Even that, however, doesn't seem to be enough to cause all the changes we feel on earth. What more could it be?

The next step is a search of the solar winds because they increase along with the sunspots. Solar winds change our upper atmosphere, and that change could trigger a chain reaction in our lower atmosphere. We know that the charged particles from solar flares are captured by the earth's magnetism because we see and measure it in the dramatic aurora borealis. And we know that air pressure changes two or three days after an aurora, sending storm systems across North America and Europe.

Most climatologists agree that the frontier of climate research lies in the solar connection. At a meeting of the National Research Council in Boulder, Colorado, in August 1982, one of the main discussions was "Solor-Terrestrial Influences on Weather." After all the reports on sunspots, solar winds, glacial sediments, and cosmic rays were discussed, the scientists decided that it would be easier to sort out the connections between solar changes and weather (if connections exist at all) if they could find physical evidence of links between the sun and the lower atmosphere.

Climate may be more sensitive to small forces in the upper atmosphere than we ever thought possible. The NRC report goes on to say that "this feeble, rarified medium cannot possibly directly influence the dense, energetic lower atmosphere where weather occurs. The answer, everyone agrees, would be a mechanism that somehow gains enough leverage to trigger changes in the lower atmosphere." But what is the mechanism? The data so far is like a pile of pieces from a puzzle. Some of the pieces fit exactly, but others fit coincidentally, or only if they're turned and adjusted just a bit. The sunspot cycles, for example, almost exactly coincide with droughts in America's Midwest. Instead of always adhering to a twenty-two-year cycle, it's more nearly an 18.6-year cycle that corresponds to a lunar cycle operating during some periods.

Dr. Douglas Paine, biometeorologist at Cornell University, says, "If there is an influence, it clearly can't be simple and straight forward. Too many people have looked too long. But we'll find it."

There is no dispute about the fact that the sun is the energy machine that drives the earth and all its systems. The NCR report challenges scientists to shift their attention from the traditional search for evidence "to a more directed effort at *understanding the physics* of the atmosphere and the solar-terrestrial system as a whole." And it adds, "Sun-weather researchers have their work cut out for them."

10

THE FOOD CONNECTION

The sun has just risen over a parched village in the Sudan. It is still cool as a few people lead scrawny cattle to the well. Soon the wind will pick up to blow sharp blasts of sand across the barren land.

Lowering their goatskin buckets into the muddy water, these people remember when they drew clear, fresh water from this deep well, when it was surrounded by forest and rich green grasses. Within the lifetime of a man, the Sudan and much of Africa has turned to desert. In the ancient Sudan trading village of Shendi, drifts of sand partly bury the houses as the boundries of the Sahara creep forward four or five miles each year.

Hillside fields in Ethiopia have eroded down to bedrock in only a few years. At rescue stations set up by the Red Cross and other charitable organizations, hundreds of thousands of starving Ethiopian children were fed at the height of the famine in 1984. Millions of dollars donated for African relief barely made a dent in repairing the damage from the worst drought of the century.

Serious soil erosion following months of drought makes agriculture impossible in many lands. (WFP/FAO PHOTO BY BANOUN/CARACCIOLO)

Africa is in trouble. Droughts across the southern border of the Sahara are part of what seems to be a natural pattern. There were three in this century. But this one has been longer and has caused more suffering.

Deserts play an important part in the climate system. Some regions get abundant rainfall simply because other regions get little or none, but arable land is turning into desert at a disastrous rate. Experts estimate that 14 million acres of farmable land vanish every year, and it's caused only partly by the years of relentless drought. Desertification is a word coined by climatologists to describe the death of arable land as deserts spread.

The lands around the ancient cities of Babylon, Ur, and Elba were once known as the "fertile crescent" because cool springs bubbled up to feed lush olive groves, and the area known as the Biblical Garden of Eden is now the parched, sun-baked Middle East. At one time, when land was no longer productive, people packed their possessions and moved on. From centuries of experience and tradition, nomads learned to leave a region before vegetation was totally destroyed so that it would have a chance to recover before they needed it again.

That simple solution isn't working now in Africa for several reasons. First, there are more people. Africa has the fastest growing population of any continent in history. Good weather or bad, Africa must feed 15 million more people each year. In all of North America, fewer than 3 million are added. In 1984, there were more than 4½ billion people on earth, and there will be twice that many by the year 2025.

"Each year the world adds the equivalent of another Mexico," writes Dr. Merle H. Jensen in a report for the University of Arizona. "How everyone will be fed is a major concern of food scientists throughout the world," a world, he reminds us, that is made up of 70 percent water, 20 percent ice and desert, and only 10 percent arable land.

Most all our food production comes from less than half of this arable land. "The tragedy is," continues Dr. Jensen, "that we are losing the finest farmlands and rich river valleys, often near cities.

It is these valleys, containing the rich soils, that fall victim to home builders, who find it easy for digging foundations; to highway builders laying down roads; and to utility companies stringing power lines."

But the number-one problem is the loss of topsoil. Wind and water wash soil away, and poor farming practices take more. Tons of soil blew away in the 1930s during the infamous Dust Bowl of America's Midwest. Drought had dried the land and burned the crops, but it was made worse by overgrazing and bad farming. That same scene is played over and over around the world, especially in Africa.

Ironically, by helping some of the Third World nations "improve" their standards of living, the industrial nations have helped destroy them. Specialists and money were sent to these countries to help dig central wells, for example, so that larger herds of cattle could be watered. But the larger herds of milling cattle stomped and trampled the grass, compacting the soil so that it continually dries and blows away. Land around water holes is overgrazed, and there are the people with more cattle to feed, with less grain, and no place to raise it in the drought-dried land.

We are tied to the land. A wealthy man living in a penthouse on top of a skyscraper is no less dependent on the land than a peasant scratching out a living on a small patch of parched soil. Except for a small amount of fish, all of the food that each American eats yearly, and all the feed for animals, is grown on 150,000 hectares of cropland and rangeland. Our lives depend upon the annual lottery of the weather. The slightest change in temperature or rainfall affects the food supply. When El Niño ravaged the fishing industry in South America, hundreds of countries felt the change in the prices and availability of fish. Climate changes that are seemingly minute can have drastic and long-lasting effects.

The Irish potato famine from 1845 through 1850 changed the course of history because thousands of people emigrated from Ireland and thousands more died in the famine. All of it began with a stretch of cold, wet weather that provided perfect conditions for a blight that destroyed that country's major crop. Holland and

In many parts of Africa people still wander from water hole to water hole with their camels, sheep, goats, and cattle. (FAO PHOTO BY A. DEFEVER)

Belgium had a similar blight, but they didn't suffer a famine because their farms didn't depend on a single crop as Ireland's did.

During the cold, unusually rainy years in Europe between 1301 and 1350, the growing season was shortened so much that grain could no longer be grown in Iceland and parts of Scandinavia. A report by Paul Waggoner in the *American Scientist* says, "This medieval example demonstrates both that the length of the season as well as the mean temperature is critical, and that the impact of climate change increases toward the poles."

The climate changes toward the equator, where drought has devastated many African nations, was thought to be a natural cycle that would straighten itself out in time. It has before, but this drought, which has been building for twenty years, has been helped along by man. What triggers a drought and what keeps it going are different factors. It may be triggered by a change in an ocean current like El Niño, but it can be sustained by overgrazing and deforestation. Thousands of acres of trees are destroyed each year for firewood to make room for more grazing land or villages.

In this cycle of continuing drought, there are three conditions that keep it going: the reduction in the amount of water held in the soil; changes in the reflectivity of the surface of the land (albedo); and changes in the supply of bacteria available to act as nuclei for the ice crystals that create rain. All of these three changes occur without man's interference, but all of them also can be triggered or speeded up by the pressures of overpopulation and land abuse.

A group of nomadic Bororo cattlemen watering zebu cattle at a new borehole in the desert, which was installed as part of a United Nations Development Program. (FAO PHOTO BY BANOUN/CARACCIOLO)

Where there is bare soil or vegetation with shallow roots, the soil loses its ability to retain water. With less moisture in the soil, there is less evaporation, so that the solar energy that would have gone into the process of evaporation goes instead into heating the air near the dry ground. With less moisture going into the air, there is less chance of rain.

Bare dry soil reflects more solar energy than dark wet soil or soil covered with vegetation. This albedo effect works to heat the near-surface air temperature and bakes the soil even more.

You wouldn't think there would be any way mankind could affect the actual formation of raindrops, but there is. Without ice crystals, there is no rain. Even in the hottest desert, rain must have at its center a nucleus of ice. Without "clean" dust there can be no ice crystals, and that's where human activity can make a difference.

The water for the ice nucleus won't freeze on just anything. Recent research has shown that an ice-crystal nucleus must be from "clean" dust from one of several species of ice-nucleating bacteria such as *Pseudomonas syringae* that live on plants.

Russell Schnell, an atmospheric biologist at NOAA, found that these bacteria produce a substance called lipoproteins. These lipoproteins provide the nucleus for ice crystals; they also allow frost to form on unprotected crops. As dead plants decay, these lipoproteins get into the soil so that particles of the organic litter also become nuclei for ice. Experiments have shown that it takes only about twenty minutes for these ice-nucleating bacteria to be carried from the soil into the clouds. NOAA teams were able to seed clouds with these bacteria to start rain.

When vegetation is stripped from the land, as in the overgrazing of the Sahel, or the Dust Bowl of America's Midwest, the ice-nucleating bacteria vanish, too.

Several teams of plant pathologists are studying the bacteria-rainfall connection to find out which plants are the best hosts for ice-nucleating bacteria. They hope to find "bioprecipitation support crops" that farmers in drought areas can plant as a kind of cloud-seeding from the ground.

Another of the man-climate connections is the greenhouse effect, which is expected to change the weather most drastically in nations between the latitudes of 35 and 45 degrees as carbon-dioxide levels rise. If the level doubles, there will be a 5.5-degree Fahrenheit (3-degree Celsius) warming across the Canadian-American border, and that will lengthen the growing season and cause less rainfall. If summer begins earlier, more moisture will evaporate from the soil, and there will be less runoff water available for irrigation.

The Colorado River shows a dramatic connection between rising temperature and available water. If there is a decrease in even a tenth of the amount of rain, along with a warming of 2.8 degrees Fahrenheit (1 degree Celsius), the Colorado's flow is reduced as much as 25 percent.

An increase in the amount of carbon dioxide in a greenhouse effect isn't all bad. Some plants will thrive. Like high-octane fuel that makes an engine work more efficiently, more carbon dioxide increases the plant's food production (photosynthesis). Experiments have shown that the nitrogen-fixing plants such as soybeans have a special advantage. Microbes in soybeans and other nitrogen-fixing plants take nitrogen from the air and turn it into nitrates, which they use for their own fertilizer. The soybeans in the study fixed 40 percent more nitrogen in air enriched with 400 parts per million of carbon dioxide than did the control plants in normal air.

Researchers look to the past as well as to the future to find how crops may be affected by warmer, drier climates. One method compares past changes in crop yields to changes in weather. The second method uses computer modeling to design programs that will simulate all kinds of combinations of temperature and precipitation changes along with what is known about how plants grow. For example, if scientists tell the computer that the temperatures in April in the Red River Valley of Minnesota will be 2.5 degrees Fahrenheit (1 degree Celsius) warmer one season, they find out that the wheat yield will increase. But if June or July of that same year is warmer by the same amount, it means more days will be hotter than 90 degrees Fahrenheit (32 degrees Celsius), and the yield will decrease.

Simulating the physical factors of the environment and the biological facts about the plants allows the prediction of future food supplies. It's a complex science. Every country is searching for ways to bring food and water to its people now and in the future. Each country has a different set of conditions and problems.

How do we solve the problems of feeding people on too little land in changing climates?

Certainly if we can begin to understand how we change the climate unintentionally, we ought to be able to figure out some ways of living with those changes.

Most scientists agree that the world must limit population growth if we expect to live as we have in the past. But we can do several other things as well. New kinds of plants can be bioengineered, or we can change old varieties to grow in a better way or under different conditions. Beaches, salt marshes, deserts, and other land never used before can be adapted for farming. And we can grow food in space.

One of the ways to use previously unused land is to change the water supply. Russia is considering ways to redirect some of her major rivers to drain wetlands and irrigate dry lands. In the United States there has been some discussion of ways to drain water from the Great Lakes and send it south and west. In February 1985, the governors of eight states and two Canadian provinces signed the Great Lakes Charter "to protect, preserve and wisely manage the single largest supply of fresh water on the North American continent." The charter was the result of reports predicting that the greenhouse effect would lower water levels through evaporation. A warming trend would extend the growing season and place a greater demand on the Great Lakes for irrigation water.

A gallon of water costs more than a gallon of gasoline in some of the states of the Persian Gulf. In Arabia they spend millions on building desalination plants to turn salt water into fresh. There has even been discussion about ways to tow icebergs from Antarctica into the Persian Gulf and tap them as a source of fresh water.

We've barely begun to discover new foods. We use about 150 of the hundreds of thousands of edible plants on earth. Three-quarters of all human food comes from just eight species of cereals, especially rice, corn, and wheat.

Dr. William Pardee, an argricultural researcher at Cornell University, tells of a new variety of corn bred for New York State growers in the 1950s. It was developed to mature in a shorter growing season than other corn. Now, in the 1980s, the growing season in New York State is even two weeks shorter. In the colder, wetter springs of the eighties, the 1950 variety of corn takes too long to sprout and mature before the first frosts, so still earlier varieties are being sought.

At Cornell, as at most agriculture colleges and research stations, new plant varieties are constantly being tailored to produce specific results—to grow in colder weather, to resist certain diseases or pests, to grow taller or shorter, with longer roots or tougher stems. But breeding a plant to do one thing does not guarantee its survival in other conditions. Thousands of acres may be sown with one variety of wheat bred to be resistant to early frost, and it may be wiped out by too much rain or drought or disease.

"Adapting a plant to climate is difficult," Dr. Pardee says as he talks about the designing of new plants by genetic engineering. "What's quoted in the press about dramatic new plants is twenty years in the future. It's not simple. It's possible, but not simple. We have to deal with twenty or thirty genes. We can move a single gene. We can grow a single cell of broccoli or rice or tobacco in a media with a toxin of a specific fungus, or salt, or perhaps a herbicide. Then we select out on the lab bench the mutant, the one that survives. But even that one may not be resistant to other effects such as temperature or soil components or disease. I have no question in my mind that it's going to be done. But to bring these things up from a single cell takes time."

Other labs are looking for ways to use old plants in new ways. Indians in America's Southwest grew tepary beans because they survive in droughts when other beans, like the pinto, do not. Tepary beans might be a good crop for dry regions now. Amaranth is another old and little-used crop. We've labeled it a weed, but it was a major food of the Aztecs. Rich in protein, with leaves like spinach and seeds that pop like popcorn, amaranth grows with a fourth of the water needed by other cereal crops.

Dr. Jensen likes to remind people that a century ago soybeans, sunflowers, and peanuts were considered unworthy of research, and now they are some of the world's most important crops.

The search is on to find ways to save space and water and still grow food cheaply. Intercropping is one answer. Long before Europeans came to America, Native Americans planted beans, squash, and corn on one hill. They may not have known the scientific reason for the plan's success, but it worked. At the Environmental Research Laboratory (ERL) at the University of Arizona, they intercrop because the beans, which are legumes, produce their own fertilizer from nitrogen in the air. Corn borrows the fertilizer from the beans, and the beans use the cornstalks for beanpoles.

In a demonstration of "intercropping" at ERL, lettuce grows on Styrofoam floats in a pool where catfish nibble on its roots and fertilize the nutrient-rich water. On an A-frame structure over the pool, melons ripen. At harvest time, the lettuce boards are removed and the melons shaken until they fall unbruised into the water, where they will float to the packing house. Tomato plants hang from moving conveyor belts with their roots dangling. The roots are sprayed with a fine mist of plant food, and the food that drips off feeds water hyacinths in the pool below. The water hyacinths can be used to produce the natural gas methane. The goal at the lab is to waste nothing, not space, not water, not even waste products.

A billion people might be fed if we could raise crops on the 20,000 miles of barren land along desert coastlines, or in tidal flats and marshlands with salty soil. Most plants can't survive in salt water. Those that can are called halophytes.

At the University of Arizona's field station in Mexico, Dr. Carl Hodges, director of ERL, and his staff are developing a halophyte crop that can be irrigated with seawater and used as forage for cattle. Pickleweed, Palmer's grass, or saltwort may some day be as common as corn on the cob or peas. The halophytes do double duty because they reclaim the land as they remove salt and mineral build-up from the soil as they grow.

Raising food in greenhouses is nothing new, unless those greenhouses are in the middle of the desert, where fresh vegetables have

(Above) *Halophytes are plants that can grow in salty soil or in salt water. They are expected to play a major role in world food production. Salt water floods this field of halophytes on a farm in Puerto Penasco, Mexico.* (Below) *In the greenhouse built in Abu Dhabi on the Persian Gulf, seawater from which the salt has been removed is used to grow vegetables in desert sand.*

(MERLE H. JENSEN, UNIVERSITY OF ARIZONA)

Sadiyat natives walk their camels past the new controlled-environment greenhouses where fresh vegetables are grown.
(MERLE H. JENSEN, UNIVERSITY OF ARIZONA)

never grown before. At Abu Dhabi in Arabia, an artificial oasis has been constructed with a tract of greenhouses full of crisp lettuce and other vegetables. It is the first such operation anywhere to be cooled with untreated seawater pumped directly from the ocean.

An orbiting space station housing its own community for months or years will need a supply of fresh food. At today's rate of about $500 per pound, it would be too expensive to ship green vegetables from the earth. It would be more practical to cultivate space gardens, and agricultural engineers are designing such farms now. One practical professor has suggested sending goats along on the space farms as the most efficient garbage disposal units because they can also provide milk and cheese, as well as fertilizer for crops.

NASA scientists are experimenting with space gardens, and so is Arizona's ERL staff, who designed most of the futuristic farms at Disney's Epcot Center in Florida, including the garden in the drum. Because plants need gravity to orient their roots and leaves, crops in space will grow in a rotating drum that creates artificial gravity with artificial light and nutrients. In the lab, spinach spins in this rotating drum with its leaves reaching away from gravity toward the light in the center of the drum, its roots flying out toward the simulated gravity.

Space farms are on the borderline of science fiction. They are possible but not yet practical. The millions of people starving in the famines of drought-stricken countries can't wait for space farms. Dr. Hodges, an atmospheric physicist who works with teams of

Spinach spins in an eight-foot plastic drum revolving around a tube of light. The roots, growing into an outer chamber that simulates gravity, receive food and water from a nutrient mist. The drum was built as part of an exhibit for the Community of Tomorrow exhibit at Epcot Center in Orlando, Florida.

(MERLE H. JENSEN, UNIVERSITY OF ARIZONA)

biologists, chemists, and nutritionists, believes that the new frontier of agricultural research lies in transforming our wastelands into farmlands and in developing plants that can survive climate changes.

"I refuse to follow the fashion of the day that predicts disaster due to world food shortage," he says. "Our only truly limited resource is time."

11

THE MAN-MADE WINTER

A shutter banged on a white frame house on Maple Street. No one fixed it. A paint-chipped rocker on the porch moved back and forth in the biting wind. No one sat in it. Lacy patterns of frost coated the bird bath and edged the rose garden. The regiment of tall maples that guarded the street stood bare and bent, stark skeletons against a bleak gray sky.

It was noon on the Fourth of July, but no one hurried to the center of town for a parade. No phones rang. No dogs barked. No children laughed. The summer picnic had been canceled by the last great climate change on earth, a nuclear winter.

"It was almost the end of the century when humanity's luck ran out," says biologist Paul R. Ehrlich in a "plausible worst-case" scenario he wrote for *Natural History* magazine in 1984. In it he described how firestorms following a nuclear blast would throw millions of tons of soot, ash, smoke, and dust from six to twelve miles into the atmosphere, shrouding the Northern Hemisphere in darkness. Choking dust would block the sunlight, and for months it would travel thousands of miles, plunging inland regions into below-zero temperatures with severe frosts. The ozone layer in the stratosphere would be destroyed by nitrogen oxides produced when thermonuclear fireballs burn the nitrogen in the air. With less ozone in the stratosphere, fewer ultraviolet rays would be filtered out, leaving plants and animals exposed to the searing rays of the sun.

A few Civil Defense officials have said that we will be able to survive the cold, barren season, which is certain to be the aftermath of a nuclear war, if only we have enough shovels. They meant to calm our fears of nuclear war, but instead they imply that we can be out of danger simply by digging holes to hide in. Even if we could dig holes deep enough to protect ourselves from radiation, we would emerge to the stark landscape of a nuclear winter. No sunlight would penetrate the dense ash hanging in the atmosphere. With no sun, green plants die, and without plants for food, animals cannot live either.

Various agencies of the federal government have offered plans for the evacuation of cities in case of nuclear attack. But even if hundreds of thousands of city dwellers could get to the country, there would still be no food, no communication, no medical supplies, and no way of stopping the nuclear winter.

Stockpiled in arsenals all over the world are an estimated 18,000 megatons of nuclear weapons. A nuclear blast of 100 megatons is enough to plunge this planet into a deadly winter. The nation firing its missles in a first strike would find that it had committed suicide. It would be a war no one could win.

Paul Ehrlich wrote that scenario to set us thinking about the future. Concerned scientists have been giving it a great deal of thought.

In spite of small groups of people who had been writing and talking about how important it is to understand what will really happen in a nuclear war, nobody paid much attention until scientists met at an international conference in 1983, where they heard

A thermonuclear explosion sends a mushroom cloud high into the stratosphere, where radioactive fallout persists for years.

(ATOMIC ENERGY COMMISSION, DEPARTMENT OF DEFENSE)

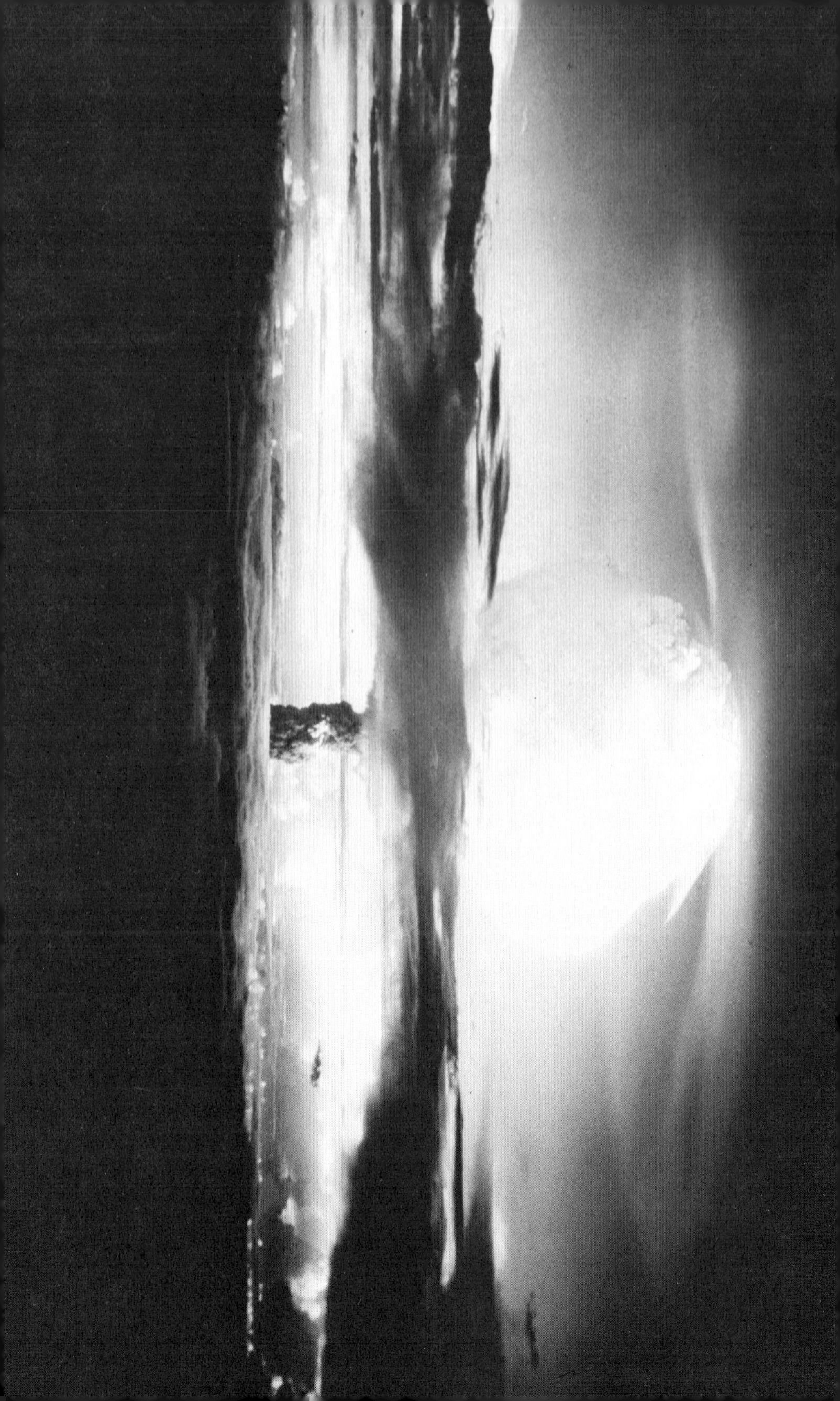

the theory of nuclear winter in a report called TTAPS. It is an acronym from the last names of its authors, Dr. Richard P. Turco, Dr. Owen B. Toon, Dr. Thomas P. Ackerman, and Dr. James B. Pollack from NASA's Ames Research Center, and Dr. Carl Sagan from Cornell University.

Using a computer model of the earth, the TTAPS group fed into it everything they knew about nuclear explosions. The computer told them that after the impact of the explosives, the greatest and most widespread danger would be from immense clouds of dust and smoke resulting from widespread fires, especially in cities. These clouds would be devastating to all life. Scientists in Russia and other European nations confirmed their findings.

The TTAPS report says, "In the aftermath of such a war vast areas of the earth could be subjected to prolonged darkness, abnormally low temperatures, violent windstorms, toxic smog, and persistent radioactive fallout—in short, the combination of conditions that has come to be known as 'nuclear winter.' " This "winter" would bring with it "the widespread breakdown of transportation systems, power grids, agricultural production, food processing, medical care, sanitation, civil services, and central government. Even in regions far from the conflict, the survivors would be imperiled by starvation, hypothermia, radiation sickness, weakening of the human immune system, epidemics, and other dire consequences. Under some circumstances, a number of biologists and ecologists contend, the extinction of many species of organisms—including the human species—is a possibility."

Before the climate modeling done by the TTAPS group, the military experts had given no thought to a nuclear winter. The Pentagon has a manual on nuclear war called "The Effects of Nuclear Weapons." It makes no mention of the effects smoke and ash might have on climate. After the bombings of Nagasaki and Hiroshima at the close of World War II, there were thousands of reports, but no studies or official notes on the raging and widespread fires that swept those Japanese cities after the first explosions.

Prodded by the TTAPS warning of fire damage, the government is going to spend 50 million dollars finding out what fires do to the

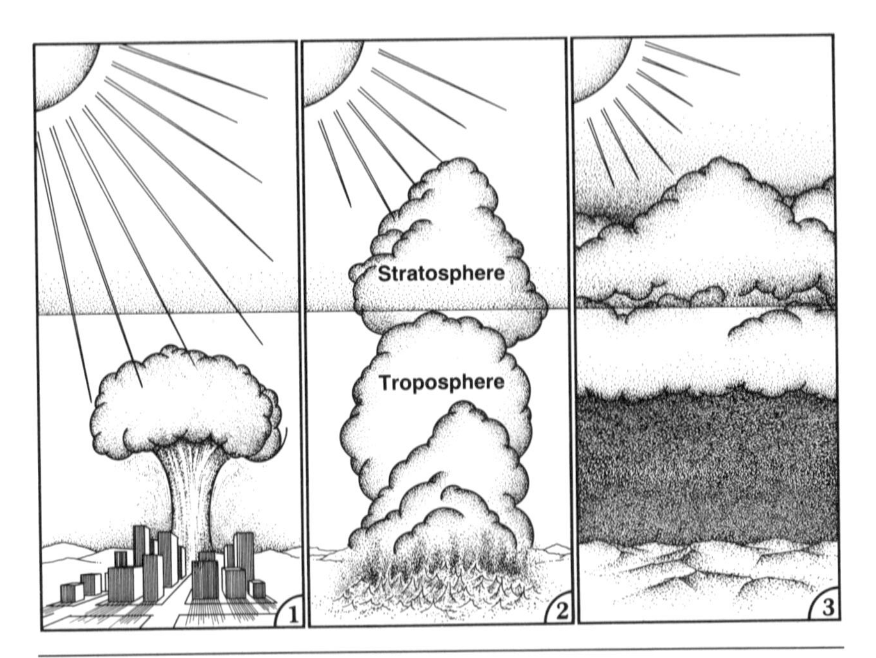

If a nuclear device explodes over a city (1), the blast will ignite fires that will merge into one enormous firestorm (2). Heat will force smoke and soot high into the troposphere, and winds will carry it around the world, blocking the sun. Rains will wash the black soot from the troposphere, but the particles of ash and debris could remain in the stratosphere for a year or more, subjecting the planet to a devastating winter (3).

atmosphere. Research teams from the Department of Energy, the National Science Foundation, the Department of Defense, the Agriculture Department's Forestry Service, and private laboratories will try to find out what happens to particles of soot and dust. How high is dust carried? How much soot is washed out by rain? Do the tiny particles of soot and smoke remain separate, or do they clump together to fall back to the earth more quickly? How fast and far do large fires spread?

If the smoke and soot studies do match the computer models showing how sunlight could be blocked enough to extinguish life on earth, dramatic changes might be made in the nation's nuclear arsenal, in the military's plans for fighting with such weapons, and in our relationships with other nations.

A nuclear winter would not be the first climate calamity on earth. Sixty-five million years ago something happened that wiped out dinosaurs along with hundreds of other species, and dozens of theories have attempted to explain what happened.

Some say the dinosaurs were too slow, too cold-blooded, or too stupid to survive, but they managed to dominate the earth for millions of years, far longer than humans have been around. In the 1960s geologists suggested that the earth itself is constantly changing. Continents and sea floors move and grind against each other, causing earthquakes, rearranging oceans, moving land masses, and triggering volcanic eruptions. They suggested that the changes in sea levels destroyed animal habitats and caused extremes in climates inland. A non-industrial greenhouse effect may have trapped the sun's energy when volcanoes loaded the atmosphere with carbon dioxide and ash.

It could have been a supernova, an exploding star that showered the earth with lethal radiation. It could have been a collision with a comet, or it could have been an asteroid hurtling into the earth. Many asteroids have orbits that cross the earth's path, and once in a while small ones do collide, leaving craters as evidence. In 1937, an asteroid named Hermes came within 400,000 miles of us, a near miss by astronomical calculations. It was a rock the size of the island of Manhattan, and if Hermes had landed in the ocean, many continents would have been awash under tremendous tidal waves. Such a collision on land would have caused earthquakes greater than the Richter scale could measure, and dust from the crash would have darkened the sky for months.

Dr. Walter Alvarez and his father, Dr. Louis Alvarez, a Nobel Prize-winning physicist, believe they have found, in a dark band of clay between layers of limestone, physical evidence for just such a collision 65 million years ago. Basing their calculations on the explosion of the volcano Krakatoa, they estimated how much dust an asteroid might have thrown into the atmosphere and how long it would have blocked sunlight. From that they agreed that plants and some marine organisms would have died first because they rely on sun for photosynthesis. And right on down the food chain, animals would have died, including the gigantic dinosaurs. They are certain that the climatic change after such a collision would have made life impossible for hundreds of species, but hundreds of new species would have taken over.

In one of the biggest controversies of science, the "what killed the dinosaurs?" research has involved astronomers, geologists, astrophysicists, biologists, and dozens of other specialists. Most agree that some kind of extraterrestrial mechanism plunged the planet into a dark era, and it was this kind of thinking that triggered the idea of a nuclear winter. If a natural calamity could shut out the sun and change the earth, then a man-made one of equal or greater impact could, too.

The search is on for Nemesis, the "death-star," as the possible cause of the "dinosaur winter." Some astronomers believe they have evidence of a companion star to the sun, one that comes close enough periodically to nudge comets out of their "natural habitat" far beyond Pluto and send them hurtling toward the earth. Most comets, which have been called "dirty snowballs," are thought to be the icy debris from a habitat called the Oort cloud some ten trillion miles from the sun. Some comets whip around the sun and back into the solar system, but some, like Halley's, return periodically. Others crash. A Defense Department satellite photograph taken in 1979 shows a comet plunging into the sun.

Researchers who disagree with the Nemesis theory are searching for a Planet X beyond Pluto. Others think that the slow, bobbing ride of the sun and the planets around the Milky Way galaxy might account for the periodical collision with comets.

Whatever caused the mass extinctions seems to return every 26 million years with the next encounter due 13 million years from now. We may never know exactly what happened. The important thing we've learned, however, is that even though the earth's dominant species could not live in the aftermath of that drastic climatic change, the earth itself did survive. It will again.

Dr. Douglas Paine, a biometeorologist at Cornell University, believes that we have to stop thinking of the earth as a spaceship controlled by man, but see it instead as a living organism. Many scientists have gone back to the old Greek word Gaia (pronounced gay-uh), which means Mother Earth. They perceive Gaia to be a vast being with man and all life as parts and partners with all its systems and cycles of ocean, air, and land. Dr. James E. Lovelock wrote a small book called *Gaia, A New Look at Life on Earth*, in which he shows how our planet functions as one self-regulated organism with the power to maintain itself as a fit and comfortable habitat for life.

Life, however, does not necessarily mean human life.

In whatever calamity destroyed the dinosaurs, Gaia was wounded, but life went on. Without the dominance of dinosaurs, mammals began to thrive. If dinosaurs hadn't disappeared, some two-legged reptile with a large brain-to-body ratio might well have developed to occupy the niche we now hold on earth. Harvard biologist Dr. Stephen Jay Gould says the catastrophe opened the way for the rise of man.

If we create a major climate change such as a nuclear winter, life on earth will change. The one-celled organisms of the ocean and land will probably adapt and go on. Paul Ehrlich suspects that burrowing animals like rats may survive along with the tough cockroach and a few other species. Mankind won't, but Gaia will. The earth's interlocked systems of ocean, air, land, and life might reel and flounder, but they'll bounce back, resettle, shove over, expand or decrease, and go on. If we create our own devastation, we may make room for some other species to dominate.

"Our deepest folly is the notion that we are in charge of the place, that we own it and can somehow run it," writes Dr. Lewis Thomas in an essay for the *New York Times Magazine*. "We are beginning to treat the earth as a sort of domesticated household pet, part park, part zoo. It is an idea we must rid ourselves of soon, for it is not so. It is the other way around. We are not separate beings. We are a living part of the earth's life, owned and operated by the earth, probably specialized for functions on its behalf that we have not yet glimpsed."

He goes on to tell us that we have an advantage over most animals because we have the kind of brain that lets us change our minds. We do not have to follow a genetically diagrammed plan for every aspect of our behavior as does the ant. The chemical messages that regulate our bodies, that are our instructions for living, tell us to think, to use our complex brains to figure out our world. " . . . we owe debts of long standing to the beings that came before us," Thomas concludes, "and to those that now surround us and will help us along in the future."

Whatever collision or explosion blocked the sun's energy from the earth 65 million years ago, it was an accident of nature. Nothing could have stopped it. But we're not dinosaurs. We don't have to wait helplessly for the destruction of a nuclear winter. We have a choice.

How will we decide to uphold our partnership with our planet? Will we be the cause of the earth's last winter?

12

WHAT'S AHEAD?

The Ice Age is coming, but the planet is getting warmer. How can that be? Who is right, those who predict a new ice age or those who tell us that icecaps will melt?

The answer is both.

Left to its natural cycle, the earth will continue to move toward a full Ice Age in the next few thousand years. Most experts agree that the unusually good weather experienced in the middle of this century is probably over and that, without interference, the climate will become colder as it moves into another Little Ice Age that seems to preview a full-scale Ice Age. A move toward a colder climate will make little difference in the Southern Hemisphere, but it will cause drastic changes in life north of the equator.

We know the earth is getting warmer. The greenhouse effect is a fact. It has already made measureable differences, and there seems to be no slowing down of the carbon-dioxide build-up. Astronomer Carl Sagan says that the scientific community is attempting to make an environmental impact statement for the entire planet on the consequences of continued burning of fossil fuels. Several facts seem certain. In a drastic man-made climate change such as a nuclear winter, all nations lose. In the warming of the planet by carbon dioxide, some lose but others gain. The U.S.S.R., for example, would benefit from a warming with ice-free ports that could expand their development of Siberia, although their grain harvests farther south would suffer. It is possible that China and other nations with rich coal supplies and a strong commitment to developing their own industries might ignore the fact that adding carbon dioxide to the atmosphere could cause destructive droughts in America or Africa.

Many climatologists believe that there is very little chance of keeping the carbon-dioxide emissions low enough to prevent the level from doubling what it was in preindustrial times and that instead of looking for a way to stop the build-up, it would be more sensible to find ways to live with it.

In a news interview, Sagan suggested three problems that need to be faced. First, we don't really know yet how severe the consequences of the global warming will be. Secondly, the solution will require sacrifices of this generation if we are to alleviate the problem for generations to come. And third, the greenhouse effect is a world-wide problem that has to be solved with world-wide cooperation. It has no boundaries.

Astrophysicist John Gribben worries that we will divert too much attention to what he calls a "technofix" and that we will gobble up resources and finance expensive research looking for ways to avoid either a global warming from the carbon dioxide or the cooling of a Little Ice Age. He echoes the concern of many scientists when he says in *Future Weather and the Greenhouse Effect* that such resources would be better used "to improve world agriculture and reduce our susceptibility to all weather variations."

Biometeorologist Douglas Paine goes so far as to suggest that man's role on earth may be to do precisely what we've been doing since the beginning of the Industrial Revolution. By adding carbon dioxide to the atmosphere, we may have stumbled on a way of warming the world and insuring our survival during the inevitable cooling of the coming Ice Age. If that's the case, then it is also our responsiblity to make sure that we develop what Carl Sagan calls a "global consciousness," a view that goes beyond this "balmy epoch" we were lucky enough to have been born into and beyond political boundaries as well.

"When we try to pick out anything by itself," wrote naturalist

John Muir, "we find it hitched to everything else in the universe."

Climate is like that. No wind blows over water that does not change weather over the land. Nothing is put into the atmosphere that does not affect life on earth in some way. And conversely, every activity of living creatures also has an impact on the blanket of air that nurtures this planet.

There are changes in the wind. Which way will the winds blow?

GLOSSARY

ACID RAIN. Rain containing particles of sulfuric acid.

ALBEDO. The ratio of solar radiation reflected compared to the amount that is absorbed. Snow and ice have an albedo of 40 to 90 percent while a forest will have about 20 percent.

ASTEROID. A small orbiting body, less than 480 miles in diameter, that revolves around the sun; also called a minor planet.

BIOSPHERE. That part of the earth's crust, water, and atmosphere that contains life.

CHLOROFLUOROCARBON. Relatively stable man-made carbon molecules that contain chlorine and fluorine. They are used as refrigerants and spray propellants. Because they absorb infrared rays, they cause a greenhouse effect when they are in the atmosphere.

CRYOSPHERE. The part of earth's surface covered by ice and snow. It is the most slowly changing component of our climate system.

DNA. Deoxyribonucleic acid, the genetic material in a cell responsible for the cell's heredity.

EL NIÑO. A warm current of water flowing from the north along the coasts of Peru and Ecuador.

FLUOROCARBONS. A class of very stable carbon compounds containing fluorine. Because they do not react with most other substances, they are useful in refrigeration, in air-conditioning systems, and as spray propellants.

GAIA. A Greek word meaning Mother Earth, now used to describe the relationship and interaction between all the physical and biological processes. In contrast to the conventional idea that life adapts to planetary conditions, this word suggests that life itself actively causes changes in land, atmosphere, and oceans.

GREENHOUSE EFFECT. A warming of the earth caused by heat trapped in the troposphere. A layer of molecules such as carbon dioxide acts as insulation. It allows the sun's energy to enter the atmosphere, but does not allow heat to leave, causing the earth to warm much as a greenhouse does.

HALOPHYTE. A plant that can tolerate salt and can grow when irrigated with salt water.

HYDROLOGIST. A scientist who studies the distribution and properties of water and its interaction between the earth and the atmosphere.

ICE CORE. A long cylinder of ice, three to five inches in diameter, removed from glaciers or icecaps containing particles that hold a record of past atmosphere.

ICE-NUCLEATING BACTERIA. These bacteria, which have some kinds of lipoprotein molecules on their surface, act as nuclei for the formation of ice crystals.

JET STREAM. A relatively narrow band of winds traveling at high speeds in a series of loops around the earth between the troposphere and stratosphere.

LIPOPROTEINS. Complex molecules composed of a protein and a fat molecule. When lipoproteins are produced by certain kinds of bacteria, they serve as nuclei for the formation of ice crystals.

LITTLE ICE AGE. A period of abnormally cold and variable weather throughout most of the world from the 1500s to the 1700s. In some parts of the world, however, this period was as early as the 1300s and extended into the early 1800s.

METHANE. CH_4, a colorless, odorless, flammable gas; also called swamp gas. It is a major component of natural gas.

MONSOON. A seasonal wind off the Indian Ocean and southern Asia that blows from the southwest in summer, bringing heavy rains. During winter it blows dry air from the north.

NITROGEN-FIXING BACTERIA. Plants such as peas, beans, and clover contain bacteria in their roots. These bacteria take free nitrogen from the air and process it into nitrates that can then be used as fertilizer.

NUCLEAR WINTER. The aftermath of a nuclear explosion leaving dust, soot, and radioactive particles in the atmosphere that would shield the earth from the sun, cause extreme cold, and endanger life.

OZONE. An explosive, poisonous blue gas, a rare form of oxygen, O_3, present in the air we breathe at only about one-thirtieth of a part per million. In the stratosphere, where it filters out much of the ultraviolet radiation from the sun, it is about five parts per million.

PERMAFROST. The layer of soil in the Arctic and Antarctic that is permanently frozen. It does not thaw during the summer.

PLANKTON. Floating plant and animal organisms in water, mostly microscopic, that are the beginning of the aquatic food chain.

SOLAR WINDS. A stream of high-energy charged particles flowing from the sun at 200 to 500 miles per second. They increase with the activity of sunspots and solar flares.

STRATOSPHERE. The layer of atmosphere that extends above the troposphere about 20 miles above the earth. It contains the belt of ozone, and the temperature increases as the altitude increases.

SUNSPOTS. Enormous areas of violent storms of charged particles passing across the sun's surface.

TRADE WINDS. A belt of winds extending 30 degrees north and south of the equator, blowing at a steady 11 to 14 miles an hour. In the Northern Hemisphere the trades blow from the east in Europe across the Atlantic toward the equator.

TRANSPIRATION. The loss of water from pores in plant leaves.

TROPOSPHERE. The bottom layer of atmosphere extending 8 to 10 miles above the equator, tapering to 5 miles at the poles. It is the layer of winds where most of our weather occurs. The higher you go in the troposphere, the colder it gets, down to minus 68 degrees Fahrenheit (minus 55 degrees Celsius).

ULTRAVIOLET. Short invisible wavelengths of solar radiation. Overexposure to ultraviolet rays can damage the eyes, cause skin to tan or burn, and cause changes in the genetic code of a cell.

BIBLIOGRAPHY

"An Acid Test for Rain." *New York Times* editorial, December 26, 1984, p. A 30.

Broecker, Wallace S. "The Ocean." *Scientific American*, Vol. 249, No. 3 (September 1983), pp. 146–160.

Brown, Lester R., and Edward Wolf. "Food Crisis in Africa." *Natural History Magazine*, Vol. 93, No. 6 (June 1984), pp. 16–20.

Brownlee, Shannon. "Forecasting: How Exact Is It?" *Discover*, Vol. 6, No. 4 (April 1985), pp. 10–16.

Bryson, Reid A., and Murray J. Thomas. *Climate of Hunger*. Madison: University of Wisconsin Press, 1977.

Canby, Thomas Y. "El Niño's Ill Wind." *National Geographic*, Vol. 165, No. 2 (February 1984), pp. 144–183.

Carbon Dioxide and Climate: The Greenhouse Effect. Hearing before the Subcommittee on Natural Resources, Agriculture Research and Environment and the Subcommittee on Investigation and Oversight of the Committee on Science and Technology, U.S. House of Representatives, 97th Congress, first session, July 31, 1981.

"The Climate Crisis." Nova script, PBS, Boston: WGBH Educational Foundation, 1983.

Conroy, Patty Ferguson. "The Greening of the Desert." *Quest 81* (March), pp. 30–35.

Critchfield, Howard J. *General Climatology*. 4th ed. Englewood Cliffs, N.J.: Prentice Hall, 1983.

Dotto, Lydia, and Harold Schiff. *The Ozone War*. New York: Doubleday and Company, 1978.

"The Effects of Pollution on Weather." *Science Challenge*. Worldwatch Institute (March 1984), p. 17–19.

Ehrlich, Paul R. "North America After the War." *Natural History*, Vol. 93, No. 3 (March 1983), pp. 4–8.

Epstein, Edward S., William M. Callicott, Daniel J. Cotter, and Harold W. Yates. "NOAA Satellite Program." *IEEE Transactions on Aerospace and Electronic Systems*, Vol. 20, No. 4 (July 1984), pp. 325–344.

Franklin, Benjamin. *The Writings of Benjamin Franklin*. Edited by A. H. Smith. Vol. IX, 1783–88. New York: Haskell House, 1970.

Gallant, Roy A. *Earth's Changing Climate*. New York: Four Winds Press, 1979.

———. *Exploring the Weather*. New York: Doubleday and Company, 1969.

Gaskell, T. F., and Martin Morris. *World Climate*. London: Thames and Hudson, 1979.

Gilford, Henry. *The New Ice Age*. New York: Franklin Watts, 1978.

Glantz, Michael H. "Floods, Fires, and Famine: Is El Niño to Blame?" *Oceanus*, Vol. 27, No. 2 (Summer 1984), pp. 14–19.

Gribben, John. *Forecasts, Famines, and Freezes*. New York: Walker and Company, 1976.

———. *Future Weather and the Greenhouse Effect*. New York: Delacorte Press, 1982.

———. *What's Wrong With Our Weather?* New York: Charles Scribner's Sons, 1979.

Herron, Michael M., Susan L. Herron, and Chester C. Langway, Jr. "Climatic Signal of Ice Melt Features in Southern Greenland." *Nature*, Vol. 293, No. 5831 (October 1, 1981), pp. 389–391.

Hodges, Carl O. "Artificial Oases." *HortScience*, Vol. 8, No. 1 (February 1973), p. 2.

Ingersoll, Andrew P. "The Atmosphere." *Scientific American*, Vol. 249, No. 3 (September 1983), pp. 162–174.

Jastrow, Robert. "The Dinosaur Massacre." *Science Digest*, Vol 91, No. 9 (September 1983), pp. 51–53.

Jensen, Merle H. *The Agricultural Challenge: Tomorrow's Food Today*. Tucson: University of Arizona, 1971.

———. "Energy Alternatives and Conservation for Greenhouses." *HortScience*, Vol. 12, No. 1 (February 1977), pp. 14–24.

Kapuza, Susan. "The Greenland Ice Sheet Program." *The Epoch*, Dept. of Geological Sciences, State University of New York at Buffalo, Spring 1982, No. 4.

Kerr, Richard A. "Mount St. Helens and a Climate Quandry." *Science*, Vol. 211, No. 4480 (January 23, 1981), pp. 371–374.

———. "Sun, Weather, and Climate: A Connection?" *Science*, Vol. 217, No. 4563 (September 3, 1982), pp. 917–919.

Lamb, H. H. *Climate, History, and the Modern World*. London: Methuen, 1982.

———. *Climate: Present, Past, and Future*. Vols. 1 and 2. London: Methuen, 1972 and 1977.

Langway, Chester C., Jr. "International Polar Research." *SUNY Research '84*, Vol. 4, No. 3 (May-June 1984), pp. 3–7.

Lovelock, J. E. *Gaia: A New Look at Life on Earth*. Oxford and New York: Oxford University Press, 1979.

Ludlum, David M. *The American Weather Book*. Boston: Houghton Mifflin, 1982.
———. *The Weather Factor*. Boston: Houghton Mifflin, 1984.
Mann, Charles. "The Weather Tamers." *Science Digest*, Vol. 91, No. 11 (November 1983), pp. 67–70.
McElroy, John R., and Stanley R. Schneider. *Earth Observations and the Polar Platform*. Washington, D.C.: National Oceanic and Atmospheric Administration, November 1984.
———. "International Cooperation in Space." Environmental Satellite, Data, and Information Service release, NOAA, January 10, 1984.
McKean, Kevin. "Hothouse Earth." *Discover*, Vol. 4, No. 12 (December 1983), pp. 99–102.
Myers, Norman. *The Primary Source, Tropical Forests and Our Future*. New York and London: W. W. Norton and Company, 1984.
National Research Council. *Solar Variability, Weather, and Climate*. Washington, D.C.: National Academy Press, 1982.
Neary, John. "Pickleweed, Palmer's Grass, and Saltwort." *Science 81*, Vol. 2, No. 5 (June 1981), pp. 39–43.
Overbye, Dennis. "Putting the Arm on Solar Max." *Discover*, Vol. 5, No. 6 (June 1984), pp. 16–21.
Paine, Douglas A. "A Climate Hypothesis Describing the Solar-Terrestrial System as a Frequency Domain with Specific Response Characteristics." *EOS*, Vol. 64, No. 26 (June 28, 1983), pp. 425–428.
"Plutonium in Glaciers." *Analytical Chemistry*, Vol. 51 (December 1979), pp. 1419A–1422A.
Projecting Future Sea Level Rise. U.S. Environmental Protection Agency, Office of Policy and Resource Management, EPA 230–09–007 revised, October 1983.
Rampino, Michael R., and Stephen Self. "The Atmospheric Effects of El Chichón." *Scientific American*, Vol. 250, No. 1 (January 1984), pp. 48–57.
Rasmusson, Eugene M., Climate Analysis Center, NOAA. "El Niño and the Southern Oscillation, An Overview." Paper read at the American Association for the Advancement of Science annual meeting, New York, May 27, 1984.
Revelle, Roger. "Carbon Dioxide and World Climate." *Scientific American*, Vol. 247, No. 2 (August 1982), pp. 35–43.
Roberts, W. O., and Henry Lansford. *The Climate Mandate*. San Francisco: W. H. Freeman, 1979.
Schneider, Stephen H., and Randi Londer. *The Coevolution of Climate and Life*. San Francisco: Sierra Club Books, 1984.
———. *The Genesis Strategy*. New York: Plenum Press, 1976.
———. "Nuclear Winter: The Storm Builds." *Science Digest*, Vol. 93, No. 1 (January 1985), pp. 48, 84.
———. "Volcanic Dust Veils and Climate: How Clear Is the Connection?" *Climatic Change*, Vol. 5 (1983), pp. 111–113.
Seidel, Stephen, and Dale Keyes. *Can We Delay a Greenhouse Warming*? United States Environmental Protection Agency, September 1983.

Stilwell, Nigel. "Our Trees Are Dying." *Science Digest*, Vol. 92, No. 9 (September 1984), pp. 39–48.

Stommel, Henry, and Elizabeth Stommel. *Volcano Weather*. Newport, R.I.: Seven Seas Press, 1983.

———. "The Year Without Summer." *Scientific American*, Vol. 240, No. 6 (June 1979), pp. 176–186.

"The Sunspot Connection." *Nova Adventures in Science*, Reading, Mass.: Addison-Wesley, 1983.

"Tales the Ice Can Tell." *Mosaic*, National Science Foundation, Vol. 9, No. 5 (September-October 1978), pp. 15–21.

Thayer, Victoria G., and Richard T. Barber. "At Sea With El Niño." *Natural History*, Vol. 93, No. 10 (October 1984), pp. 4–12.

Thomas, Lewis. "Man's Role on Earth." *New York Times Magazine*, April 1, 1984, pp. 36, 37.

Thompson, Kenneth. "Forests and Climate Changes in America: Some Early Views." *Climatic Change*, Vol. 3 (1980), pp. 47–64.

Thompson, Louis M. "Our Changing Climate." *Science of Food and Agriculture*, Vol. 1, No. 2 (May 1983), pp. 2–8.

Turco, Richard P., Owen B. Toon, Thomas P. Ackerman, James B. Pollack, and Carl Sagan. "The Climatic Effects of Nuclear Winter." *Scientific American*, Vol. 251, No. 2 (August 1984), pp. 33–43.

Waggoner, Paul E. "Agriculture and Carbon Dioxide." *American Scientist*, Vol. 72 (March-April 1984), pp. 179–182.

Webster, Bayard. "Forests Role in Weather Documented in Amazon." *New York Times*, July 5, 1983, pp. C7–8.

What Is the Greenhouse Effect? Washington, D.C.: U.S. Environmental Protection Agency, 1983.

Woodwell, George M. "The Carbon Dioxide Question." *Scientific American*, Vol. 238, No. 1 (January 1978), pp. 34–43.

World Food Report 1984. Rome: Food and Agriculture Organization of the United Nations.

Young, Louise B. *Earth's Aura*. New York: Alfred A. Knopf, 1977.

INDEX

Numbers in italics refer to illustrations.